Springer Series in Reliability Engineering

Today's modern systems have become increasingly complex to design and build, while the demand for reliability and cost effective development continues. Reliability is one of the most important attributes in all these systems, including aerospace applications, real-time control, medical applications, defense systems, human decision-making, and home-security products. Growing international competition has increased the need for all designers, managers, practitioners, scientists and engineers to ensure a level of reliability of their product before release at the lowest cost. The interest in reliability has been growing in recent years and this trend will continue during the next decade and beyond.

The Springer Series in Reliability Engineering publishes books, monographs and edited volumes in important areas of current theoretical research development in reliability and in areas that attempt to bridge the gap between theory and application in areas of interest to practitioners in industry, laboratories, business, and government.

Indexed in Scopus and EI Compendex

Interested authors should contact the series editor, Hoang Pham, Department of Industrial and Systems Engineering, Rutgers University, Piscataway, NJ 08854, USA. Email: hopham@rci.rutgers.edu, or Anthony Doyle, Executive Editor, Springer, London. Email: anthony.doyle@springer.com.

More information about this series at https://link.springer.com/bookseries/6917

Adam Jabłoński · Marek Jabłoński

Digital Safety in Railway Transport—Aspects of Management and Technology

Springer

Adam Jabłoński
Management and Quality Institute
WSB University in Poznań
Poznań, Poland

Marek Jabłoński
Management and Quality Institute
WSB University in Poznań
Poznań, Poland

Translated by
Barbara Borkowska-Kępska
Dąbrowa Górnicza, Poland

Andrew Gillin
Katowice, Poland

ISSN 1614-7839 ISSN 2196-999X (electronic)
Springer Series in Reliability Engineering
ISBN 978-3-030-96132-9 ISBN 978-3-030-96133-6 (eBook)
https://doi.org/10.1007/978-3-030-96133-6

Translation and Language Services: Barbara Borkowska-Kępska, Andrew Gillin

This Springer imprint is published by the registered company Springer Nature Switzerland AG
The registered company address is: Gewerbestrasse 11, 6330 Cham, Switzerland

Introduction

In the context of dynamic technological progress and the resulting development of the digital transformation processes of railway enterprises, there is a need to clarify this issue in a broad way to understand cyber security mechanisms as a key challenge for the rail transport sector. Railway enterprises which are responsible for operation and maintenance processes and the safe movement of people and goods must be aware of the increased risks arising from the digitalization of all their areas of operation. Dynamic digital transformation processes include both the elements of the configuration of the business models of railway enterprises and the ontologically separated railway system itself, understood as the railway network and railway vehicles intended for traffic on that network, distinguished by the functional and technical characteristics. Both areas overlap in many cases by creating a complex set of related elements, where railway traffic safety management poses a major challenge to the digitalization of the areas of operation of railway enterprises. Railway enterprises create mechanisms to manage railway traffic safety by introducing management systems based on the concept of the management of the risks occurring in their operation. The division of risk of rail market operators is due to both their place and role in the railway system. The type of activity determines the extent to which the operation of a railway enterprise using a given business model has an impact on the level of railway traffic safety. The digitalization of different areas of rail activity, as a widespread phenomenon induced by both the needs of its own development and legal conditions, should ensure that all processes directly or indirectly linked to its implementation and operation are safe. Due to the high level of innovation in the digital economy, it is difficult to avoid any doubt that the railway system under the conditions of digital transformation will continue to be as safe as before digitalization. Many doubts in this respect arise from the lack of extensive operational experience. Standardized digital safety management mechanisms should therefore be sought, based on the conscious role of the business model of railway enterprises and the embedment of this problem in the context of management, social and technological conditions. The social nature of railway safety stems from strong public pressure to ensure high levels of railway traffic safety. A condition for maintaining trust in rail transport is to invest in the improvement of increasingly better risk control measures built up not only in the railway system itself, but also in the configuration of increasingly used digital business models of railway enterprises. This problem is considered not only terms of scientific research, but also its practical application, which makes it a valuable solution for stakeholders in the rail transport sector as well as professionals dealing with management processes geared to different sectoral circumstances. The focus on management, social and technological aspects makes this topic clear from many points of view, which should make readers aware of the importance of digital safety in the rail transport sector and from how many perspectives the methods to improve its management processes can be sought. The interdisciplinarity of this subject means that the scope of critical evaluation of this issue requires an extremely broad view, as shown in this monograph. This work is one of the key results of many years of work by the authors in the field of safety management in rail transport, digital business models and the social aspects of designing digital business models.

The main objective of this scientific discussion is to present, in a multidimensional and holistic way, the mechanisms of digital rail transport safety in management, social and technological terms. The monograph also takes the role of sustainable business model design into account by focusing research and analysis on identifying the role of safety as a social factor. The structure of the book consists of 13 chapters, each of which begins with the introduction and ends with the final conclusions. The work ends with conclusions and empirical-methodical conclusions. Chapter 1 is methodological, describing the scope and methodology of research into digital safety management mechanisms as part of the social business models of railway enterprises. Chapter 2 presents the original models of the authors in terms of different views on the concept of designing and defining safety management mechanisms in rail transport, with an explanation of areas such as the mechanisms of effective safety management in rail transport, the mechanisms of efficient management of railway sidings, the mechanisms of shaping safety culture in rail transport, the mechanisms of ensuring technical compatibility and safe integration in rail transport and the mechanisms of ensuring safe operation of railway vehicles. Such a wide scope of analysis helps to better understand the complex context of the issue of the rail transport sector in the context of the safety criterion. Chapter 3 deals with the clarification of the issue of the digital business models of railway enterprises, with an indication of the social aspect, including the assumptions for sustainable business models and sustainable supply chains in rail transport. All this has been presented in the context of the process of digitalization of rail transport. Of particular importance in this chapter is the reliance on theoretical discussion and the establishment of the leading theory in which research and analysis are embedded, namely the theory of cybernetics, which is the basis for a systemic approach and the starting point for the development of the theory of digitalization. Scientific achievements in this area are extremely helpful in explaining numerous contemporary phenomena pertaining to the analytics of complex digital technical systems. Chapter 4 deals with the typical sectoral determinants of rail transport in the context of issues relevant to the objective and subject of the book, such as the place and role of railway market operators in terms of sharing responsibility for safety, clarifying the essence of the railway system and its technical, operational and organizational conditions. Issues of technological management were also indicated in terms of improving the performance of the technical railway system and safety criteria, as well as information management in the rail transport chain, and railway traffic safety. Attention was also paid to the assumptions of the operation of control and traffic management systems in terms of safety, the safe operation of railway vehicles and the place and role of railway traffic safety management systems, all of which was important in order to clarify the conditions for the operation of the railway system in terms of railway traffic safety criteria. Chapter 5 introduces readers to the understanding of the digital transformation processes in rail transport, sets the determinants of digital transformation and defines the directions and areas of digital transformation of rail transport. It also draws attention to the assessment of the digital maturity of railway systems in technical, operational and organizational terms and to the migration of responsibility in hybrid, i.e. analog-digital, railway systems. Chapter 6 deals with important issues for digital safety, i.e. it clarifies digital safety in rail transport by highlighting issues such as theoretical assumptions for defining digital safety in rail transport in holistic terms, cybersecurity in rail transport and technical system safety—safety and security aspects, the life cycle of railway systems and the digital safety of technical systems, as well as the identification, design and installation of safety criteria in railway hardware and software solutions. All these issues justify the claim that this problem is difficult and requires extensive knowledge of the rail transport sector. Chapter 7 is devoted to identifying the place and role of processes for creating digital innovation in rail transport. It defines content to understand innovation in the context of rail transport specificity, problems and barriers to the use of innovative solutions in rail transport, technical, operational and organizational innovations and safety in rail transport. This chapter also addresses important issues and problems in testing, verifying and validating innovative digital solutions in the context of safety criteria, and clarifies innovation issues in relation to the exchange of information in railway systems in subjective and objective terms. Chapter 8 aims to identify modern trends and emerging technologies which increase the level of safety in rail transport. They include solutions such as artificial intelligence and its application in improving railway safety, simulations, computer games, videos and virtual reality (VR) and railway traffic safety management, the Internet of Things and interface management in the process of improving railway traffic safety, predictive maintenance of railway infrastructure and railway vehicles in terms of safety, Big Data and its assumptions in shaping digital solutions in rail transport, Industry 4.0 and 5.0 and their assumptions in shaping digital solutions in rail transport, a digital twin as a concept for improving railway traffic safety, and finally digital solutions in rail transport in the field of communication systems in railway traffic control. Chapter 9 is devoted to an important issue, which is currently being reconstructed in relation to the dynamic digital transformation processes of railway enterprises affecting their safety culture, concerning the development of safety culture in rail transport through the creation and implementation of digital solutions. The chapter indicates that railway enterprises should be understood as organizations that meet the definition of high reliability organizations. The following part of this chapter explains the organizational culture of railway enterprises in relation to digital transformation processes, the determinants of shaping safety culture in railway enterprises towards digital solutions, and a model of shaping safety culture through the digital transformation of railway enterprises, which is important for the content of this chapter. Chapter 10 deals with an important element of safety management in rail transport. This chapter seeks to explain the determinants of railway safety monitoring processes using digital economy solutions. Theoretical and practical assumptions for railway safety monitoring, methods and techniques for monitoring railway safety, digital solutions for monitoring railway safety, the model for digital monitoring of rail transport safety, as well as an important element for ensuring the safety of railway traffic controlled by digital devices, namely data transmission systems as an aspect requiring monitoring due to its role in the data and information safety management system, as well as the control of critical tasks for railway safety have been identified. Chapter 11 presents the findings of the monograph. It outlines the conceptualization and operationalization of the digital safety model for rail transport. The chapter identifies digital technologies supporting the development of a digital safety model for rail transport, safety management methods and tools supporting the development of a safety model for digital rail transport, as well as the safety model for digital rail transport itself. The digital safety model is a coherent set of devices, technical systems and components supporting a safety management system using digital solutions as a basis for maintaining the expected level of safety. The digital safety model is a combination of digital economy solutions embedded in individual components—the functions of the railway system. It is such a set of elements shaping the configuration of the railway system, whose protection against defects, failures, incidents, disasters, and any other irregularities are built on a cybersecurity canvas based on technical, legal-organizational and standard solutions. The twelfth and final chapter presents the original safety triad in the business models of rail transport enterprises, described in terms of management, digitalization and socialization. This chapter presents the results of research and analysis in the form of a description of the conditions of synthesis of the simultaneous implementation of the analog and digital layer of safety management mechanisms in rail transport, a hybrid model of rail transport safety management based on the mechanisms of analog and digital layers, and the conceptualization of the safety triad in the business models of railway transport enterprises. The triad consists of a set of three interconnected elements, namely safety management mechanisms as a set of rules, procedures and methods used to achieve the expected level of railway safety. The second element is the digitalization of processes related to rail transport activities, with a particular focus on digital safety. The final element of this triad is the socialization of the concept of railway safety, which arises from the expectations of stakeholders in terms of intolerance for railway incidents. The book ends with general and empirical and methodical conclusions. A broad review of the literature, the observation of railway enterprises in this area and the concepts and conclusions developed by the authors confirmed the importance of the issue of digital safety in rail transport in social and technological terms. Digital safety issues are also of particular importance when set against the background of digital business models. This topic is extremely broad, not only in terms of the holistic nature of the safety issue, but in particular the specificity of the operation of the rail transport sector. In this sector, in terms of two key ontologically separable research spaces, namely safety and security, the decision was made to combine them into a single model called Cyber Operational Safety (COS), the assumptions of which should cover all risks arising from the use of computer applications directly related to safety and those that support operational, maintenance and logistics processes in rail transport. The final conclusions of the work on the monograph neatly summarize the above-mentioned issues. The book should inspire other researchers interested in digital safety management mechanisms and socially oriented digital business models.

Contents

About the Authors

Adam Jabłoński is an Associate Professor at WSB University in Poznań, Faculty in Chorzów, e-mail: adam.jablonski@ottima-plus.com.pl. He is also President of the Board of "Ottima Plus" Ltd. of Katowice, a reputable management consulting company, and Vice-President of the "Southern Railway Cluster" Association of Katowice, which supports development in railway transport and the transfer of innovation, as well as cooperation with European railway clusters (as a member of the European Railway Clusters Initiative). He holds a postdoctoral degree in Economic Sciences, specializing in Management Science. Having worked as a management consultant since 1997, he has gained experience and expertise through co-operation with a number of leading companies in Poland and abroad. He is the author of a variety of studies and business analyses on business models, value management, risk management, the balanced scorecard and corporate social responsibility. He has also written and co-written several monographs and over 100 scientific articles in the field of management and the railway sector. Adam's academic interests focus on the issues of modern and efficient business model design, digital business models, including sustainable business models, and safety management systems in the railway sector.

Marek Jabłoński is an Associate Professor at WSB University in Poznań, Faculty in Chorzów, Poland, e-mail: marek.jablonski@ottima-plus.com.pl. He is also Vice-President of the Board of "Ottima Plus" Ltd. of Katowice, a reputable management consulting company, and President of the "Southern Railway Cluster" Association in Katowice which supports development in railway transport and the transfer of innovation, as well as cooperation with European railway clusters (as a member of the European Railway Cluster Initiative). He holds a postdoctoral degree in Economic Science, specializing in Management Science in the following topics: performance management and business models. Having worked as a management consultant since 1997, he is the author of a variety of studies and business analysis on business models and safety in railway systems, implementing solutions for high reliability organizations. He has also written and co-written several monographs and over 100 scientific articles in the field of business models and safety aspects of the railway sector. Marek's academic interests focus on the issues of innovative digital business models, safety management in railway systems and new trends in these areas.

List of Figures

Chapter 1
Scope and Methodology of Research into Digital Safety Management Mechanisms as a Social Element of the Business Models of Railway Enterprises

Adam Jabłoński and Marek Jabłoński

Abstract *Purpose*: The purpose of the chapter is to present the key scientific and application assumptions of the scientific problem of digital safety management in the context of the determinants of the business models of railway enterprises

Design/Methodology/Approach: A critical process of analysis and the conceptualization of research has been used as the methodology for this chapter, aiming to identify the key determinants of the development of the digital safety issue. A traditional approach has been used to define theoretical and utilitarian goals in management, social and technological terms.

Findings: The research assumption requires a broad review of the relevant literature, case studies and an analysis of legal, management, social and technological solutions. The research is interdisciplinary.

Research limitations/implications: The limitations in the scope of the analytical and evaluation study result from the very wide range of analyses conducted and the impossibility of broadly analyzing the various types of solutions in the digital economy. The focus was on the main determinants of the issue.

Practical implications: In addition to theoretical assumptions, this chapter identifies a number of analytical needs in the field of utilitarian research.

Originality/Value: The originality of the analytical and research intention lies in the combined concept of considering the concept of digital safety set against the background of the issue of business models of railway enterprises in terms of management, social and technological factors.

1.1 General Overview of Scientific Discussion

The issue of digital safety in rail transport in the context of the determinants of digital business models of railway enterprises is currently a very dynamically developing area of science. In both theoretical and practical terms, it is interdisciplinary and

A. Jabłoński and M. Jabłoński, *Digital Safety in Railway Transport—Aspects of Management and Technology*, Springer Series in Reliability Engineering,
https://doi.org/10.1007/978-3-030-96133-6_1

combines the attributes of management, technology and the digital economy. From this perspective, safety is one of the building blocks of sustainable business models based on sustainable supply chains. The digital transformation determines a new perspective on safety in rail transport, which needs to be clarified and embedded in the paradigm of creating social values. This should be seen in this way as the safety of the railway system is a factor awaited by a wide group of stakeholders in society.

From such a cognitive perspective, the authors wish to present the issue of the digital safety of rail transport in management, social and technological terms.

1.2 Theoretical Assumptions

Main objective of scientific discussion

The main objective of this scientific discussion is to present, in a multidimensional and holistic way, the mechanisms of rail transport digital safety in management, social and technological terms.

Research methodology and the scope of scientific research and discussion

This monograph is one of the results of many years of work by the authors in the field of safety management in rail transport, digital business models and the social aspects of designing digital business models.

In order to achieve the research objective, qualitative longitudinal research into railway enterprises and companies that rely on digital business models has been used. Scientific research in this area has been published in the cycle of scholarly monographs (Jabłoński and Jabłoński 2018, 2019, a, b, c, 2021a; b, c) on rail transport safety management mechanisms, business models and socially oriented digitalization focused on social aspects. The sequence of research is confirmed by the order of published scholarly monographs, where parallel studies were conducted on safety management in rail transport and on the aspects of digital business models, including the identification of social factors in this area. This scientific cycle related to the scientific interests of authors also includes this monograph, extending the scientific achievements of its authors in some areas and in a certain sense representing the culmination thereof at this point in time.

The subject of scientific discussion

The subject of scientific discussion includes scientific reflections, theoretical and application studies on the digitalization of the rail transport sector and the resulting implications for rail transport safety. In this case, a critical literature review and a study of practical solutions were conducted.

At the same time, keywords were defined during the critical literature review and were used to select and review scientific articles as a source of theory. The selection of literature and practical solutions focused on scientific publications and solutions mainly from the last ten years published in world-renowned scholarly journals.

Cognitive goals

Cognitive goals include:

1. Analysis of the theoretical assumptions of the mechanisms of creating digital safety in the business models of railway enterprises.
2. Analysis of the theoretical assumptions of identifying the relationship between the digital determinants of the business models of railway enterprises and digital safety management.
3. Analysis of digital transformation process in rail transport and its impact on safety management processes, treating safety as an important social factor.
4. Analysis of digital innovation management processes in rail transport in terms of improving rail transport safety.
5. Analysis of the impact of digitalization on the development of safety culture in rail transport.

Methodological goals

Methodological goals include:

1. Discussing key issues related to digital safety, its definitions, assumptions and practical implications in the rail transport sector.
2. Identifying digital safety management mechanisms in the context of the digital transformation of the business models of railway enterprises and railway organizational, social and technical systems.
3. Developing the conceptualization and operationalization of the digital safety model of rail transport.
4. Determining whether there are sufficient grounds for creating a new paradigm of digital safety management in rail transport.

Utilitarian goals

Utilitarian goals include:

1. Presenting the mechanisms of management and improvement of modern railway digital safety to managers.
2. Presenting the key mechanisms of the implementation of modern and innovative solutions of the digital economy and its impact on the expected level of safety of railway systems.
3. Presenting the assumptions for improving rail transport digital safety by shaping the business models of companies, treating safety as a key social factor for the managers of railway enterprises.

Key research questions

The following research questions were posed within the defined research objectives:

Question 1. Does the digital transformation of the business models of railway enterprises and systems have an impact on railway traffic safety?

Question 2. What digital safety management mechanisms determine technological progress?

Question 3. What are the key problems faced by rail transport safety managers in implementing innovative solutions for the digital economy?

Question 4. What are the key factors that determine aspects of cybersecurity in rail transport?

Question 5. What are the limitations of the different technologies and solutions of the digital economy in the context of rail transport safety?

Question 6. What determinants influence the treatment of rail transport safety as a social factor?

Embedding the subject in the theory of management sciences, safety theory and the theory of technical sciences:
Due to the multidimensionality and holistic nature of the functioning of rail transport, the subject of scientific discussion requires addressing issues related to the theory of management sciences, safety theory and the theory of technical sciences.

A cognitive gap:
In order to identify a cognitive gap, the relevant literature and databases on the research problem were reviewed. This was also based on the experience resulting from the long-term process of studying management, social and technological processes in the rail transport sector. Longitudinal research was used for this purpose. The cognitive gap is an insufficient recognition of the processes of shaping the expected level of railway safety through management, social and technological actions resulting from the dynamic digital transformation of both the railway system as such and the digitalization of the business models of railway enterprises. It is also necessary to recognize the assumption that rail transport safety is an important social factor due to the expectations of different stakeholder groups. It then influences the development of sustainable business models of railway enterprises and a sustainable supply chain. Digitalization means that the approach to safety management in rail transport must take a number of new determinants requiring accurate identification, recognition, analysis and scientific research into account.

This book is based on the following assumptions:
Macroeconomic assumptions:

1. Given the globalized rail market and the related liberalization thereof, railway systems must meet expectations of increased efficiency and safety.
2. The development of the digital economy shapes a new perspective on the interpretation of legal, organizational and technical solutions for improving interoperability procedures and the safety management processes of railway systems.

Sectoral assumptions:

1. The place and role of the rail sector in the digital economy is crucial and requires the implications of many new legal, technical and organizational solutions.
2. The railway sector is becoming one of the key elements in the development of the green objectives of the global economy.

3. The railway sector must take into account the principles of interoperability and a high level of safety.
4. The railway sector and its digital transformation affect other sectors of the economy, stimulating them by means of growth opportunities.
5. Rail transport safety is desirable as a social value.

Microeconomic assumptions of railway enterprises:

1. Railway enterprises should be strongly focused on a constructive comparison of digital transformation processes and ensuring the expected level of railway safety.
2. Railway enterprises and their business models should be a source of permanent innovation in digital solutions that should help improve rail transport safety.
3. Railway enterprises should build efficient digital business models with a focus on understanding safety as a strong social factor.

The topicality and importance of the issue and the contribution of the issues related to digital safety in terms of management and technological conditions to science

By writing this book, the authors wished primarily to contribute an original and comprehensive approach related to defining the mechanisms of effective safety management in rail transport to the development of the theory and practice of digital safety management in rail transport. The literature has not so far conducted such extensive scientific discussion on digital safety, either in subjective terms (the digital determinants of the business models of railway enterprises) or in objective terms (the determinants of the implications of digital innovation in rail transport).

The clarity of the creative contribution is related to the presentation of a comprehensive scientific discussion and practical solutions in the field of rail transport digital safety.

The effectiveness of the creative contribution is related to achieving the utilitarian goals of the monographs, i.e. presenting multidimensional strategic recommendations for the development and improvement of digital safety in rail transport to managers and decision-makers. The developmental character of the creative contribution is that it will provide the basis for further research and scientific discussion in this area.

References

Jabłoński A, Jabłoński M (2018) Mechanizmy efektywnego zarządzania bezpieczeństwem w transporcie kolejowym. CeDeWu, Warszawa

Jabłoński A, Jabłoński M (2019) Modele biznesu przedsiębiorstw, Perspektywy Rozwoju – ujęcie koncepcyjne. CeDeWu, Warszawa

Jabłoński A, Jabłoński M (2020a) Mechanizmy efektywnego zarządzania bocznicami kolejowymi. CeDeWu, Warszawa

Jabłoński A, Jabłoński M (2020b) Mechanizmy kształtowania kultury bezpieczeństwa w transporcie kolejowym - czynnik ludzki i organizacyjny. CeDeWu, Warszawa

Jabłoński A, Jabłoński M (2020c) Social business models in the Digital Economy. New Concepts and Contemporary Challenges. Palgrave Macmillan, Switzerland
Jabłoński A, Jabłoński M (2021a) Digital business models, perspective on monetization. Routledge, New York
Jabłoński A, Jabłoński M (2021b) Mechanizmy zapewnienia zgodności technicznej i bezpiecznej integracji w transporcie kolejowym. CeDeWu, Warszawa
Jabłoński A, Jabłoński M (2021c) Railway safety culture model, theory and practical implications. CeDeWu, Warsaw

Chapter 2
Safety Management Mechanisms in Rail Transport

Adam Jabłoński and Marek Jabłoński

Abstract *Purpose:* The purpose of this chapter is to present key assumptions for identifying safety management mechanisms in rail transport.

Design/Methodology/Approach: A critical literature review has been adopted as a methodology for this chapter in the context of defining the achievement of the authors in terms of different thematic and methodological approaches to defining effective safety management mechanisms in rail transport.

Findings: This chapter identifies key approaches to identifying and defining the various models of safety management in rail transport.

Research limitations/implications: The limitations of the analytical and evaluation study result from the interdisciplinarity of the analyses, which makes the scope of analyses and studies very broad.

Practical implications: The practical nature of the issues discussed has been demonstrated in the resultant nature of the proposed methods and approaches to safety management in rail transport.

Originality/Value: The originality of the analytical and research intention is the comprehensive identification of various concepts for defining safety management mechanisms in rail transport.

2.1 Introduction

One of the key elements used to ensure an acceptable level of safety in rail transport is the definition of such logical descriptions and accepted principles that are repeatable and standardized. These descriptions and principles should set out interlinked procedures which are easy to implement and understood by all employees in railway enterprises. Their interpretation should determine dynamic decision making by the right people, generating long-term and short-term objectives within the set criteria for the effectiveness and efficiency of entities in the rail transport sector. Appropriate safety management mechanisms in rail transport are to be used for this purpose. The purpose of the adoption of these mechanisms is to achieve specifically measurable results in the area of safety, according to the "zero accidents" principle. This principle

A. Jabłoński and M. Jabłoński, *Digital Safety in Railway Transport—Aspects of Management and Technology*, Springer Series in Reliability Engineering,
https://doi.org/10.1007/978-3-030-96133-6_2

should guide any adopted solution, procedure, instructions, method of conduct or the assumptions of structuring an organization in the context of organizational, management, technological and digital assumptions. Only the adoption of a systemic and procedural methodology can then determine the above-mentioned principles for the design and implementation of effective and efficient safety management mechanisms in rail transport.

2.2 The Mechanisms of Efficient Safety Management in Rail Transport

A proper understanding of the logic of the functioning of the rail transport system in terms of safety arises primarily from certain assumptions and adopted solutions which ensure its smooth operation, which generates useful value for the system. From this perspective, the concept and requirements for adopted mechanisms, i.e. the descriptions of operation methods or the order of phenomena in safety management processes in rail transport, emerges. These mechanisms are determined by the context of a rail organization, adopted and understood by all stakeholders, i.e. the so-called high reliability organization. An attempt to clarify rail transport safety management mechanisms is based on a process approach aimed at identifying the crucial procedural rules that must be fulfilled so that the expected level of safety of a social and technical system, such as the railway system, is ensured.

Efficiency as an economic concept defines the relationship between the costs incurred in maintaining the safety of a railway system and the results in this respect. Safety, by one definition, means freedom from unacceptable risk of harm (Commission Implementing Regulation (EU) No 402/2013). The actions of railway enterprises in this area should serve to mitigate risks until their level of acceptance is reached. In technical systems, a highly improbable risk means an occurrence of failure at a frequency less than or equal to 10^{-9} per hour of operation (Commission Implementing Regulation (EU) 2015/1136). This should be the way technical systems are designed and they should achieve this level of probability after life cycle changes have been made. Safety management mechanisms in rail transport are designed to ensure that the technical systems in operation maintain the expected level of safety. This is particularly important because the railway system is composed of a large number of different components, which together form a unique configuration, which must be safe at all times. The level of reliability of individual components, as well as of the whole system, must be consistently high.

The maintenance systems of railway infrastructure and railway vehicles must be highly effective and efficient, leading to low failure rates. This is what safety management mechanisms, which are procedurally built in the safety management systems implemented by infrastructure managers and railway undertakings, are used for. Such an approach characterizes the railway system as a highly complex system. Therefore, a challenge faced by the rail sector is the ability of rail market participants to combine management, technological and operational aspects into one coherent system. A modern approach to safety issues puts emphasis on the aspects of improving safety management systems and maintenance management systems towards the dynamic development of the rail sector in the context of systems, processes and models. The theory of rail transport safety management is important if we wish to understand this issue in the context of other sciences, as this is indispensable to explaining both theory and application solutions. Figure 2.1 presents the theory of rail transport safety management set against the background of management science and safety theory.

The original approach to shaping and understanding the new theory of rail transport safety management reveals the interdisciplinarity of this issue, in which it should be assumed that the theory of rail transport safety management is first of all embedded in the area of safety management theory, and more broadly, in management science theory. An important part of this theory is safety theory in technical and operational systems, which is part of safety theory. In addition, issues covered by the theory of rail transport safety management science include, to a lesser extent, other theories of science, e.g. psychology (human factor), ergonomics (work environment) and others (Jabłoński and Jabłoński 2018). Safety management mechanisms are therefore

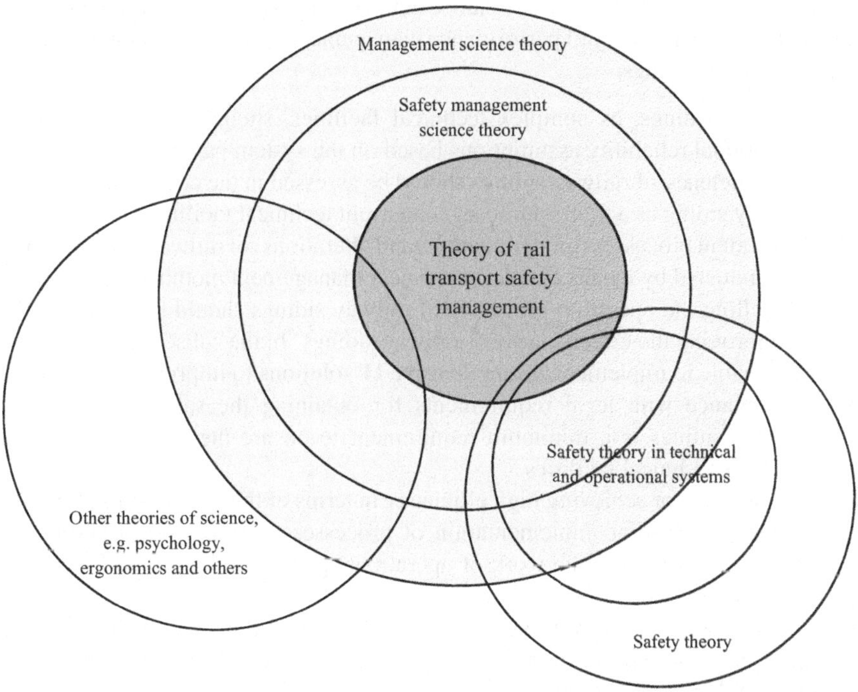

Fig. 2.1 New theory of rail transport safety management set against the background of management science and safety theory (*Source* Jabłoński A, Jabłoński M (2018) Mechanizmy efektywnego zarządzania bezpieczeństwem w transporcie kolejowym [Mechanisms of effective rail transport safety management]. CeDeWu, Warsaw, p. 334)

interdisciplinary and comprise a set of good practices that should be continuously reviewed in the digital transformation process.

2.3 The Mechanisms of the Efficient Management of Railway Sidings

Railway sidings are a key link in the functioning of the rail transport system. They are the source of the shipment and collection of cargoes. They are used for maneuvering, handling and repair work. The larger the number of efficient railway sidings, the better the functionality of the railway system, which affects both the accessibility and safety of rail transport. Railway sidings improve railway traffic and enable key operational activities to be conducted. It is vital to treat railway sidings as complex technical and operational systems, the role of which is to support railway traffic and increase the attractiveness of the rail transport sector. On the basis of analyses devoted to the operation of railway sidings in terms of the efficiency of the railway system and safety, the following strategic recommendations for railway siding managers in terms of improving their efficiency can be defined, which at the same time constitute assumptions for the theory of performance management of complex technical objects such as railway sidings:

1. Railway sidings, as complex technical facilities, should be managed using operational reliability assumptions based on the system paradigm.
2. The efficiency of railway sidings should be assessed in the context of treating a railway siding as a multi-purpose, concurrent technical facility.
3. Investment processes (modernizations and alterations) of railway sidings should be conducted by means of relevant project management methodologies.
4. Modelling the operation processes of railway sidings should be a crucial step in improving the effectiveness of railway sidings. In the subsequent stage, it is reasonable to implement comprehensive IT solutions to improve their work.
5. Compliance with legal requirements for obtaining the safety certificate of railway sidings is a minimum requirement to ensure the suitability of such complex technical facilities.
6. A condition for achieving high efficiency in terms of the functioning of railway sidings is the joint implementation of processes to ensure the usability and good organization of the work of operational processes, measured by specific performance indicators.
7. The efficiency of railway sidings will determine the ability of the siding to perform the requested services as expected.
8. The implementation of processes increasing the efficiency of the railway siding should be preceded by an analysis of the business model of the functioning of the railway siding (legal and technical and organizational aspects) (Jabłoński and Jabłoński 2020a).

High efficiency is the sum of the efficiency of individual processes and projects, aimed at improving logistics operations and investment projects aimed at increasing the efficiency of the organization of the work of railway sidings. High efficiency depends to a very large extent on the level of usability of the railway siding, understood as a complex technical facility. The efficiency of a complex technical facility will strongly depend on the extent of the operational use of the facility.

2.4 The Mechanisms of Shaping Safety Culture in Rail Transport

Safety culture is a condition for maintaining the focus on the constant maintenance of high reliability of the functioning of the railway system. The efficiency of the technical system depends on the approach of managers to the management of railway enterprises in terms of safety. Safety is not constant: it always depends on dynamic changes in its management process, the resources involved, the efficiency and reliability of technical measures, and organizational and operational processes. Risks should be under constant supervision, and safety-oriented leadership is indispensable. The process of shaping safety culture in rail transport is complex. Building a positive climate and safety culture in railway enterprises requires a huge effort by their managers and the full commitment of staff at all levels of management. The mechanisms of shaping a positive safety culture will be all technical and organizational concepts, methods, techniques and solutions used to shape such a safety culture and enable its identification and evaluation (Jabłoński and Jabłoński 2021b, p. 146). Figure 2.2 shows the model of shaping safety culture in rail transport. It indicates a number of solutions which, when used jointly in railway enterprises, should contribute to the development of a positive climate and safety culture.

The presented model consists of several key interconnected components. On the left side of the model, the extremely important role of the leadership of the management of companies and their managerial intentions are indicated. They give stakeholders an impulse to act according to the established rules using a system of values and a high level of responsibility. Risk and uncertainty management takes into account a working environment that is relevant to recognizing railway enterprises as high reliability organizations. The environment is complex, unsafe and uncertain; the enterprise, being aware of this, conducts a process of looking for errors to avoid disasters. In addition, it is necessary to ensure a balance between mindfulness and mindlessness. Additionally, it detects anomalies, introduces alerts and responds to them. The safety management system is applied and improved by the enterprise taking into account the context of the organization, which corresponds to the business model, the strategy applied, processes and projects. The technologies and technical systems used play an important role. The key role in shaping safety culture is taking human and organizational factors into account as a priority. Within this area of knowledge and activities, the leading role is played by the issues

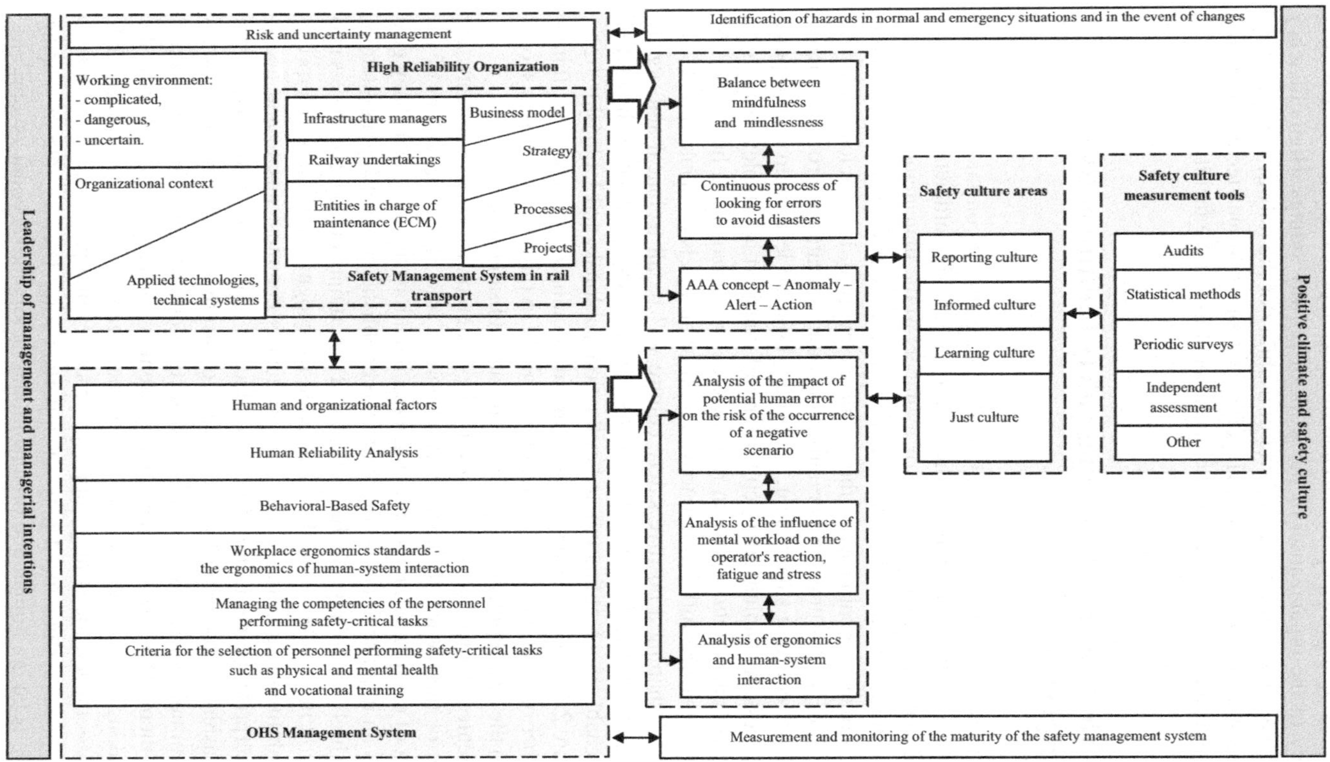

Fig. 2.2 Model of shaping safety culture in rail transport (*Source* Jabłoński A, Jabłoński M (2020b) Mechanizmy kształtowania kultury bezpieczeństwa w transporcie kolejowym - Czynnik ludzki i organizacyjny [The mechanisms of shaping safety culture in rail transport — the human and organizational factor]. CeDeWu, Warsaw, p. 440)

of human reliability analysis, behavioral-based safety, and workplace ergonomics standards, including the ergonomics of human-system interaction. Here it is important to ensure the expected competencies of personnel performing safety–critical tasks, and to provide criteria for the selection of such personnel, taking areas such as physical and mental health and vocational training into account. It is important to analyze the impact of potential human errors on the risk of a negative scenario, and to analyze the impact of mental workload on the operator's reaction as well as his or her fatigue and stress. Additionally, the ergonomics of human-system interaction should be analyzed. The issues of human and organizational factors should also be covered by the requirements of the OHS management system standards. Within the framework of such defined areas of railway traffic safety management, the principles of employee behavior in the context of the safety culture being shaped should be implemented. Rules for reporting occurrences, understood as a reporting culture, and rules for providing information about safety issues, understood as an informed culture, should be developed. In addition, the principles of the learning culture and just culture should be defined. In order to manage resources effectively and efficiently for the needs of shaping safety culture and to achieve ambitious goals in this regard, safety culture should be measured and monitored. As part of measuring safety culture, periodic safety culture audits should be conducted, statistical methods should be used if necessary, periodic survey questionnaires should be conducted to evaluate the implementation processes of the mechanisms used, and the independent assessments of the safety culture should be commissioned. As part of the process of identifying hazards in normal and emergency situations and in the event of changes, changes to the procedures should be made when necessary. The measurement and monitoring of the maturity of the safety management system should be an ongoing process. Based on the presented methodology of conduct for the needs of shaping safety culture in rail transport, railway enterprises should strive to create the expected positive climate and safety culture in rail transport (Jabłoński and Jabłoński 2020b, pp. 441–442). A positive safety culture will be a condition for obtaining the expected behavior of managers and staff at all levels of management which is necessary to maintain a high level of railway traffic safety.

2.5 The Mechanisms of Ensuring Technical Compatibility and Safe Integration in Rail Transport

Experience in proving technical compatibility and, in particular, safe integration has not taken into account the need for a structured and systemic approach, which explains the ontological boundaries for both concepts. First of all, technical compatibility concerns the assessment of the correct application of technical solutions, whereas safe integration more broadly requires technical systems to be verified. Technical compatibility is verified by Notified Bodies (NoBo) to assess conformity with the TSI, and Designated Bodies (DeBo) for compliance with national regulations and

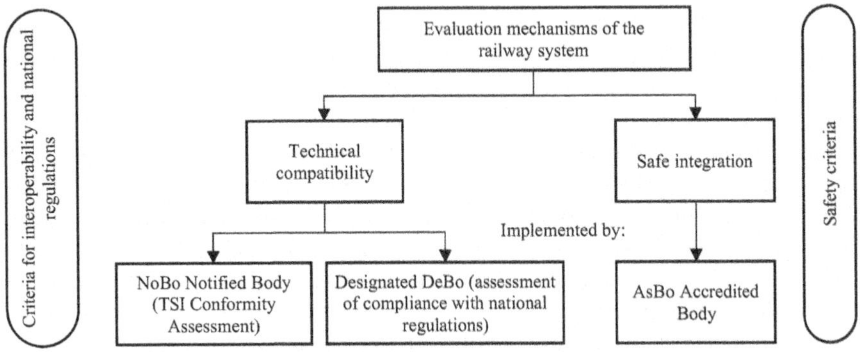

Fig. 2.3 Mechanisms of assessing the railway system (*Source* Jabłoński A, Jabłoński M (2021a) Mechanizmy zapewnienia zgodności technicznej i bezpiecznej integracji w transporcie kolejowym [Mechanisms of ensuring technical compatibility and safe integration in rail transport]. CeDeWu, Warsaw, p. 20)

safe integration by Assessment Bodies (AsBo). Figure 2.3 shows the mechanisms of assessing the railway system, including technical compatibility and safe integration.

From this perspective, safety should be assessed throughout the lifecycle of the product with regard to the design, construction, operation and decommissioning process. Safe integration will first apply if the railway system is changed simultaneously from the technical and functional perspective.

The primary objective of engaging in the process of technical compatibility verification and safe integration is to ensure the expected level of safety of railway traffic. To this end, a number of measures should be conducted referring to different theories derived from a systemic approach. Firstly, the architecture of rail technical systems should be identified, recognized and shaped in order to maintain their coherence throughout the railway system. In order to ensure the integration of these railway systems, the interconnection programs should be managed. In doing so, attention should be paid to the features and role of self-adaptive technical systems in railway systems, as they may pose a challenge for modern railway technologies. The starting point in the context of complex technical systems is the theory of Systems of Systems, as it is at the forefront of ensuring the integration of technical systems. Resistance to disturbances and ways of protecting technical systems against hazards poses a major challenge in terms of an engineering approach to technical compatibility and safe integration in order to ensure a high level of standards of safe management of railway traffic and rail transport preceded by the processes of admission to operation, renewal and modernization of railway infrastructure. The effective management of technical, operational and organizational change in rail transport is a condition and the result of the verification of technical compatibility and safe integration. The standardization of interfaces between different types of equipment as part of the configuration of railway systems in investment processes, including the design, construction, renewal and modernization of railway systems, will be a crucial problem to be solved in both

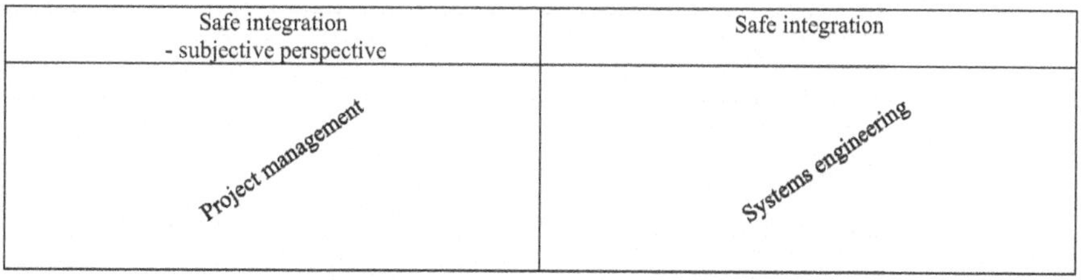

Fig. 2.4 A model for ensuring safe integration in rail transport in the context of a change project in the railway system (*Source* Jabłoński A, Jabłoński M (2021a) Mechanizmy zapewnienia zgodności technicznej i bezpiecznej integracji w transporcie kolejowym [Mechanisms of ensuring technical compatibility and safe integration in rail transport]. CeDeWu, Warsaw, p. 311)

the technical and organizational context of the design approach. Effective and efficient management of interfaces will be a consequence of their proper standardization. As a result, risk management, including the valuation thereof, will be necessary at all stages of the conformity assessment and verification processes. The system thus configured should ensure that the verification processes of technical compatibility and the safe integration of the life-cycle of railway systems are conducted correctly. This theoretical concept of technical compatibility and safe integration should be operationalized in the form of a utilitarian methodology, i.e. easily implemented in the conditions of strategic and operational activity of railway enterprises.

Figure 2.4 shows the original application model for ensuring safe integration in rail transport in the context of a change project in the railway system.

The proposal presented consists of two areas for which safe integration should be conducted, namely at the level of the project management process (subjective perspective) and at the technical level (objective perspective) based on the fulfilment of the objectives of systems engineering. From the organizational and management perspective, design interfaces should be identified and assessed in terms of risk and be reviewed, verified and validated (Jabłoński and Jabłoński 2014, pp. 65–72), whereby these activities should confirm that any solution at project management level is appropriate and in line with contractual and legal arrangements. The second part of the process of ensuring safe integration should be based on a technical approach, whereby all technical interfaces constituting communication protocols between equipment that is interconnected within the railway system cooperate correctly and do not generate an unacceptable level of risk leading to unexpected situations. Both concepts are presented in detail in Fig. 2.5.

The presented diagram indicates the scope of system integration through the use of safe interfaces. These activities aim to ensure safe integration in terms of project and systemic approaches. In the context of the project approach, seven thematic areas have been identified which require the identification of safe interfaces in order to ensure safe integration. These areas include: different suppliers of systems/products/railway components, different installation times, various testing requirements, the adoption of multiple project management methodologies, different requirements for design and

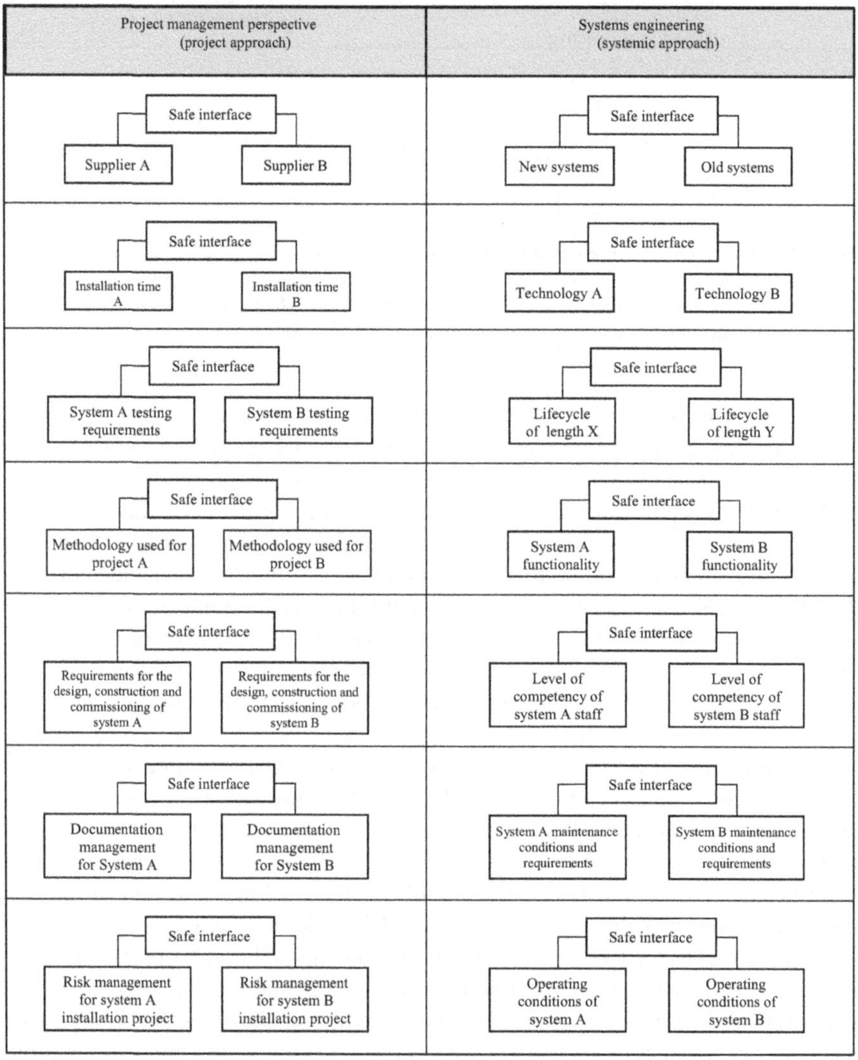

Fig. 2.5 Scope of system integration through the use of safe interfaces to ensure safe integration in terms of the project approach and systemic approach (*Source* Jabłoński A, and Jabłoński M (2021a) Mechanizmy zapewnienia zgodności technicznej i bezpiecznej integracji w transporcie kolejowym [Mechanisms of ensuring technical compatibility and safe integration in rail transport]. CeDeWu, Warsaw, p. 313)

construction in terms of placing in service, different ways of documentation management, and different ways of project risk management. All these areas may hinder the process of safe integration of railway systems/products/components in terms of the project management process. In terms of engineering, when connecting railway systems/products/components, safe integration will be affected by their different systems, technologies, life cycle stages, functionalities, levels of competence of operating staff, maintenance conditions, and operating conditions (Jabłoński and Jabłoński 2021a). The complexity of the issue of technical compatibility and safe integration in rail transport forces designers, users and experts to conduct a holistic and interdisciplinary assessment based on the analysis of the technical system life-cycle. A safe system is a system which is technically compatible at all times on the railway network, demonstrates the expected efficiency and is safely integrated with other elements of the railway system.

2.6 The Mechanisms of Ensuring the Safe Operation of Railway Vehicles

A crucial aspect of rail transport safety management is ensuring the safe operation of railway vehicles. In terms of the life cycle of a railway vehicle, the concept of safety requirement capture in the process of placing a vehicle in service will also be important. Another important aspect will be the implementation of a safety-oriented system for introducing changes to existing types of railway vehicles in the context of legal conditions. A significant number of problems may arise in the management of the configuration of railway vehicles in the context of changes in the life cycle. Finally, the identification and management of railway safety–critical elements will be worth considering in terms of the operation of railway vehicles. All these issues relate to the implementation of the railway vehicle maintenance management systems, required for Entities in Charge of Maintenance (ECMs), including if they perform such functions, railway undertakings and infrastructure managers. In addition, rolling stock repair workshops can implement these systems on a voluntary basis. These systems are based on ensuring a high level of operational readiness and safety. As mentioned, one of the most important aspects of maintaining the safety of railway vehicles in operation is the proper configuration management, which means a systemic organizational, technical and administrative process covering the entire life cycle of a vehicle or vehicle type. This is conducted to ensure and maintain consistency in terms of documentation and traceability of changes, in order to ensure the following: compliance with applicable laws, control of changes and their documentation in the technical documentation or in the documentation accompanying the authorization issued, and the validity and accuracy of the information and data stored, so that the relevant parties receive notification of the changes where appropriate. In terms of the safe operation of railway vehicles, it is essential to ensure a proper maintenance process of the railway vehicle throughout its life cycle and to monitor the impact of parts replacement on the configuration of the railway vehicle in terms of safety. It may then be necessary to upgrade or renew the railway vehicle. Four options for the impact of the replacement of railway vehicle parts on determining the status of changes in the configuration of the vehicle can be distinguished, namely:

- Maintenance replacement / modifications that do not change the technical documentation,
- Modifications which make changes to the technical documentation, which require verification but do not affect the essential design features,
- Renewal or modernization - changes which alter the essential design features but do not require authorization,
- Renewal or modernization which alters the essential design features and requires new authorization.

Therefore, in accordance with the overall objective of managing and mitigating the identified risk to an acceptable level, an applicant for placing the vehicle in service or one introducing substantial changes to a vehicle type, before submitting an application to the authority issuing the relevant administrative decisions, implements a process of defining the applicable requirements, within which all the necessary requirements for the design of the vehicle for its life cycle must be properly identified, assigned to functions or subsystems or are addressed through conditions for use or other restrictions and implemented and validated. The essential requirements for subsystems according to the technical specifications for interoperability (TSI), the technical compatibility of the subsystems within the vehicle and the safe integration of the subsystems within the vehicle, and the technical compatibility of the vehicle with the network in the area of use (Commission Implementing Regulation (EU) 2018/545) must be fulfilled. The digitalization of maintenance processes is important for the operation of railway vehicles and related safety management mechanisms. Predictive maintenance of railway vehicles, which consists of the optimal use of vehicles, is one of the maintenance strategies. It is based on anticipating the future condition of the facilities and taking appropriate corrective, maintenance and preventive actions in a timely manner. This is enabled by the continuous monitoring of their technical condition and constant monitoring of the desired parameters. For this purpose, various types of detectors are used, which cooperate with IT systems to monitor the technical condition of individual components of railway vehicles. Prediction allows for the determination of future states of objects thanks to continuous measurement (including in real time), as well as the analysis of the values of current and historical states, and estimation of future values. As a result, the optimal planning of maintenance, repair, replacement and other works is possible, which minimizes the number of occurrences of failures and allows for the achievement of longer intervals between failures. By taking action at the most favorable time, this method turns out to be more effective in preventing and resolving failures than the method of preventive maintenance. Safety–critical components that require particular attention and priority in maintenance procedures should also be analyzed and monitored by means of digital solutions. The critical aspects of each component are always related to the specific design of the vehicle and the specific functions of the component. Essential parts of safety–critical components should be defined for safety management purposes. This aspect of the safe operation of railway vehicles should be supported by relevant IT systems, the functionalities of which should include the monitoring of vehicle configuration - the register and status of part replacement, identification and recording of defects, identification of hazards, risk categorization and risk management, and the determination of criticality of railway vehicle components based on risk analysis. All these essential functions should be integrated into a digital management system for the configuration of railway vehicles and critical elements for railway safety. The mechanisms of the safe operation of railway vehicles must be based on these thematic areas. In terms of the management of railway safety–critical elements, it is necessary to:

- Identify all the components that can be separated as a stand-alone system from the main components constituting the configuration of the whole vehicle, together with an indication of the sub-components,
- Separate within each sub-component indicating safety impact (S), or operational readiness (R),
- Select from the sub-components only those that have an impact on safety (S),
- Develop, for each component/sub-components affecting safety (S), the FMEA analysis indicating the Priority Risk Number,
- Implement the results of the FMEA analysis in the hazard register, supplement/correct and indicate, in the documentation of the maintenance system, priority in maintenance procedures for components/sub-components which are critical to railway safety,
- If it is appropriate, amend the documentation of the maintenance system with regard to maintenance procedures,
- Monitor these activities within the framework of a dedicated IT system.

This system of railway vehicle maintenance management focused on configuration management processes and critical elements for railway traffic safety means that the most important risks of loss of safety characteristics are subject to special supervision. The mechanisms thus designed, supported by predictive maintenance systems, should ensure the expected level of safety of the railway vehicle operation and maintenance processes.

2.7 Conclusion

The starting point for understanding the specificity of different safety approaches is to define rail transport safety management mechanisms as well as the mechanisms of efficient management of complex technical facilities such as railway sidings, the mechanisms of shaping safety culture as well as technical compatibility, safe integration and the safe operation of railway vehicles. This is necessary to indicate the context and interpretation of railway traffic safety in rail transport. A broad and somewhat unified approach creates the conditions for a proper interpretation of this issue. Mechanisms describing the mode of action or the sequence of phenomena provide good practice in the context of certain conditions pertaining to the context of the functioning of a particular ontological entity. It can be an enterprise, a technical facility, a complex technical system, etc. This chapter focuses on proving that a correct and standardized action aimed at improving railway traffic safety must be based on clearly described operating mechanisms. Only they can protect people from making mistakes and mitigate the risks associated with the lack of proper responses to these risks. Management and acting mechanisms set standards, rules and procedures. The discussion in this chapter confirms that the concept of searching for standardized mechanisms works in the process of improving railway traffic safety, while setting out best practices for managing railway traffic safety.

References

Commission Implementing Regulation (EU) No 402/2013 of 30 April 2013 on the common safety method for risk evaluation and assessment and repealing Regulation (EC) No 352/2009

Commission Implementing Regulation (EU) No 2015/1136 of 13 July 2015 amending Implementing Regulation (EU) No 402/2013 on the common safety method for risk evaluation and assessment

Commission Implementing Regulation (EU) No 2018/545 of 4 April 2018 establishing practical arrangements for the railway vehicle authorisation and railway vehicle type authorisation process pursuant to Directive (EU) 2016/797 of the European Parliament and of the Council

Jabłoński A, Jabłoński M (2014) Walidacja stosowania wspólnych metod oceny bezpieczeństwa w transporcie kolejowym w warunkach polskich z uwzględnieniem zarządzania konfiguracją. Technika Transportu Szynowego 1–2:65–72

Jabłoński A, Jabłoński M (2018) Mechanizmy efektywnego zarządzania bezpieczeństwem w transporcie kolejowym. CeDeWu, Warszawa

Jabłoński A, Jabłoński M (2020a) Mechanizmy efektywnego zarządzania bocznicami kolejowymi. CeDeWu, Warszawa

Jabłoński A, Jabłoński M (2020b) Mechanizmy kształtowania kultury bezpieczeństwa w transporcie kolejowym - Czynnik ludzki i organizacyjny. CeDeWu, Warszawa

Jabłoński A, Jabłoński M (2021a) Mechanizmy zapewnienia zgodności technicznej i bezpiecznej integracji w transporcie kolejowym. CeDeWu, Warszawa

Jabłoński A, Jabłoński M (2021b) Railway safety culture model, theory and practical implications. CeDeWu, Warsaw

Chapter 3
Digital Business Models and the Rail Transport System

Marek Jabłoński

Abstract *Purpose*: The aim of this chapter is to present crucial assumptions for defining theoretical and practical assumptions on digital business models in the context of the specificity of the rail transport sector.

Design/Methodology/Approach: The methodology for this chapter is a critical literature review, the authors' theoretical and practical reflections on the theory of cybernetics and its impact on the development of the digitalization of the rail transport system. Attention was paid to the aspects of sustainable business models in terms of the digital transformation of rail transport.

Findings: This chapter identifies key approaches to understanding the place and role of digitalization in shaping the socio-technological systems of rail transport.

Research limitations/Implications: The limitations of the analytical and evaluation study are due to the interdisciplinarity of the analyses, which makes the scope of analyses and studies very broad. A large number of issues are being addressed in this thematic area in order to achieve a high level of consistency in inferences.

Practical implications: The practical nature of the issues discussed has been demonstrated in the resultant nature of the proposed methods and approaches to attempts to explain the processes of rail transport sector digitalization in the context of the concept of business models.

Originality/Value: The originality of the analytical and research intention is the comprehensive identification of various approaches to defining the theoretical foundations for understanding the specificity of digital business models against the background of cybernetics. It was also important to clarify that rail transport enterprises shape social business models through their role in the sustainable supply chain.

3.1 Introduction

Nowadays, railways in Europe and around the world are changing very dynamically, and the dynamics of this change has surprised almost everyone. This is due to the fact that, in principle, most of the technological, digital and organizational and management solutions currently being developed are directly applicable to rail transport.

A. Jabłoński and M. Jabłoński, *Digital Safety in Railway Transport—Aspects of Management and Technology*, Springer Series in Reliability Engineering,
https://doi.org/10.1007/978-3-030-96133-6_3

However, this is particularly true for the digitalization of the rail transport sector. Technological conditions and social pressure towards high-quality and safe railways are undoubtedly conducive to the permanent search for the best possible solutions and economically optimal in the field of railway traffic control, railway infrastructure management, railway traffic management or the maintenance of railway vehicles and other vehicles. According to this interpretation, digital railways will be the driving force behind the integration of a comprehensive digital mobility system for all modes of transport (Wei and Li 2019). The integration of transport is becoming crucial in relation to digital solutions that will at the same time have a universal dimension, which is relevant for a given sector or mode of transport. In order to understand the assumptions for the construction of digital railways, it is necessary to find the origin and initial input necessary for the creation of creative methods, techniques, tools, and accelerator platforms for the dynamic growth of innovative creations applicable to rail transport. They lie in the logic of the adopted principles of numerous theories, from complexity theory, the theory of technical systems and their architecture to the theory of cybernetics and many others. Their combined implementation and use can guarantee a well-run and justified digital transformation of rail transport in Europe and throughout the world.

3.2 The Theory of Cybernetics in the Digitalization Process of the Rail Transport Sector

One of the crucial theories that are particularly important in the digitalization process of the rail transport sector is the theory of cybernetics, understood as a transdisciplinary approach to control systems and the related transfer of information (communication) in humans and machines. According to Norbert Wiener, widely considered the founder of cybernetics, classic cybernetics is the science of the general laws of control and information, transfer in machines, living organisms and society. He suggested dividing all parts of the unit into two parts to ensure stability and balance. These two parts can be simply referred to as the subject and object of management, among which vertical relationships are developed (Wiener 1948).

Cybernetics is an interdisciplinary science which was created at the junction of mathematics, logic, semiotics, physiology, biology and sociology. Its inherent features include the analysis and revelation of general principles and approaches in scientific cognition (Novikov 2016, p. 7). Cybernetics is based on the idea of interdisciplinary science. As such a science, cybernetics is the science of control. The cybernetic automatic regulation system can be called a system designed to solve new regulation tasks, which are more complex than classical stabilization tasks, as well as a program control and tracking system. Feedback in cybernetic systems uses a wide range of logical elements and often works in a state of constant search (oscillation). Issues related to the control and generation, transformation and reception of control signals (Iwachnienko 1962, pp. 21–22) play a central role in cyber systems.

In the logic of cybernetics in relation to any system, one of two attitudes can be adopted:

- either leave the system alone and just watch it to learn as much about it as possible, and then we will have cognitive problems to solve;
- or turn the system into another system, and then we will have decision-making problems to solve.

In other words, we may either interfere or not interfere with the system, and because it is logically complete, there are not—and cannot be—problems other than from the two listed groups (Mazur 1976, p. 90). The determination of the essence of classical cybernetics is based on the ideas of classical scientific rationality (Lepsky 2015). Within cybernetics theory, there are first, second and third-order cybernetics.

First-order cybernetics includes:

- the general laws of the control processes and transmission of information;
- observable systems,
- allopoietic machines,
- classic scientific rationality,
- the "subject -object" paradigm.

Second-order cybernetics includes:

- cybernetics observing systems,
- the response to trends in the object of management,
- triplet integration-training-adaptation,
- homeostasis support and self-organization.
- Third-order cybernetics includes:
- the self-developing polysubjective (reflective-active) environment,
- attention to the intellectual component,
- the empirical generalization of large data volumes,
- Power, Culture and Institutions.

Fourth-order cybernetics is also being developed: it includes independent robotic complexes (RC)—fourth-order full autonomy, self-reproduction (Chepin 2021).

P. Asaro believes that there are three interpretations of cybernetics:

1. A narrow view, i.e. science researching feedback control,
2. A broad view, i.e. "cybernetics is all, and we live in the Century of Cybernetics",
3. An internal (epistemological) view, i.e. second-order cybernetics (pressure on the feedback of the controlled system and observer) (Asaro 2010, pp. 169–186).

The applications of cybernetics are common in computer science, especially in the fields of artificial intelligence, neural networks and control engineering. However, some new features have been formulated in these cybernetics principles. The new cybernetics emphasizes autonomy, self-organization, understanding and the role of the observer in modelling the system (Yang et al. 2017). The behavioral data of users is an important indicator for product improvement. Today's web applications collect various types of service data, which are the ideal source of information for product designers to better understand the needs and behavior of users (Liu et al. 2017).

In rail transport, both theoretical and applied cybernetics can be used. Theoretical cybernetics examines the general principles of systems operation without referring to specific physical objects. It was created at the junction of biology and mathematics, drawing on many separate areas such as:

- mathematical logic,
- control theory, regulation theory,
- systems theory,
- information theory,
- game theory,
- decision theory,
- the theory of abstract machines,
- praxeology,
- neural network theory.

Applied cybernetics deals with the practical application of the general laws of theoretical cybernetics. Its branches that are applicable to rail transport include:

- technical cybernetics (technocybernetics)—the automatic control of machines and technological processes,
- economic cybernetics—the automation and optimization of diagnostics, planning and management,
- social cybernetics (sociocybernetics)—the use of cybernetics rights in social sciences.

Cyber-based rail transport systems are based on control processes. Two basic systems can be distinguished in the control processes, namely:

1. A controlling system, sometimes called a process organizer.
2. A controlled system, also called a control object.

Control relationships occur between these systems consisting of the deliberate impact of the controlling system on the controlled object in order to achieve the desired changes in the course of controlled processes. The object and a controlling system form a control system (Mynarski 1981, p. 94).

Autonomous systems occur in rail transport, and disturbances may occur therein which are a source of railway incidents: rail disasters, major accidents and accidents. An autonomous system is any system in which there is a set of feedback aiming at functional equilibrium, regardless of how such a system was created (by construction, birth or coincidence). The consequence of an autonomous system to maintain the ability to control is to counteract factors that could lead to the destruction of the system, thus extending the system's lifetime. The autonomous system can counteract destructive factors in two ways. The first is to prevent changes in the system's surroundings which may cause an imbalance to the system's functional equilibrium, and the second is to eliminate disturbances if they occur. In the first of these possibilities, the system does not allow for the creation of danger by combating its causes (prevention), while in the second it fights the effects of the danger that has already arisen (therapy) (Mazur 1966, p. 57). For example, in rail transport, the concept of the Internet of Things (IoT) is strongly developed in its essence, which uses, inter alia, system and cybernetics theories.

It is vital to understand the concept of the Internet of Things (IoT) and the dangers of over-simplification, which can lead to failures in terms of complex systems/systems of systems (SoS). Therefore, this requires an exploration of what information science and cybernetics can teach us in terms of complexity control in its many forms (Schuck 2021). Given the modern understanding of cybernetics applicable to railways, it is necessary to understand how to arrange a large number of various models of parts of the brain and at many different levels of detail, bearing in mind that it is important to look for many basic principles, examples of which are different combinations, to present an increasingly extensive landscape, ranging from the various aspects of cybernetics to brain theory and cognitive science with AI insights to robotics (Arbib 2018). However, one should be aware that despite random attempts to update cybernetics theory, current technology has gone beyond this theory, and current tools and models do not explain and predict how people interact with modern interfaces (Mulder et al. 2016). With the emergence of new computer paradigms, software systems must be automatic, self-adaptive and evolutionary. These dynamic features make the software behavior uncontrolled, which can result in unwanted interactions and system disruption. To solve this problem, the concept of software cybernetics is now being introduced, which tries to control the development process of the software using intelligent feedback (Chen et al. 2017). An important dilemma of cybernetics is to show how cybernetics brings together various knowledge sets of different stakeholders, including professional evaluators, resulting in coherent knowledge (Levin-Rozalis 2010).

3.3 Digital Business Models in Rail Transport

A business model describes how the organization creates, delivers and captures value (Osterwalder and Pigneur 2010). In digital terms, it is important to move from e-business to digital business as the presence and integration of things, connected and intelligent, with people and business. This kind of connection requires a network that connects everything together and a cloud infrastructure that provides all the services needed to cope with every requirement of business agility nowadays (Lopez 2014).

The prefix 'cyber' regularly introduces new terms, namely a cybersystem, cyberspace, a cyber threat, cybersecurity, etc. (Novikov 2016, p. 107). Digitalization significantly contributes to increasingly efficient and reliable railway systems and companies operating in the rail market using digital business models. The introduction of new technologies such as 5G, IoT and cloud computing will be another important step towards improving the efficiency and effectiveness of railway systems. This allows for the collection and analysis of Big Data in real time, the optimization of the system and increased smart automation, which underlies the shaping of digital business models of companies which provide services in the rail transport sector.

The use of the cloud platform in modelling digital business models offers a significant advantage, namely that various entities in the rail transport sector, which differ significantly in tasks or size and structure, can cooperate on an equal basis and maintain the same level of safety. In this context, rail transport service providers use all information available via the cloud to optimize railway processes. Digitalization is therefore an indispensable driving force for modernizing the entire rail transport system and related business models. Developed and co-created new business models pose a particular challenge in terms of combining their functionality with the railway system as a complex system of devices, equipment, technical systems and work organization. Only the concurrent implementation of the assumptions of digital business models with railway system interfaces will create synergies between the two areas, and the boundaries between them will disappear. It should be noted that legal regulations in the rail transport sector restrict the digital exchange of data, as analog documentation is required, as evidenced by the current state of knowledge and experience. However, it is necessary to think from a long-term perspective and note that fundamental efforts, research and projects to digitize these processes are being made in this respect. Therefore, railway systems based on digital solutions and the digital business models of railway enterprises should be considered a target unity, which should be a challenge to achieve the full digitalization of railways. Such safety will have to be considered in terms of a new paradigm based on the assumptions of the digital economy. Innovative business models should ensure data integration in terms of their functionality, supported by an in-depth analysis of technological and logistical processes and an approach to serving the potential market and customers. Digital solutions combining the digital transformation of railway systems with digital business models are very often supported by so-called digital twins, which makes it possible to reconstruct and simulate railway processes in the context of technological progress management at the intersection of the railway system and the business model. Access to data from digital twins allows for the design and modeling of new efficient and sustainable processes using cloud-based solutions. Modern digital business models of companies operating in the rail transport ecosystem can be designed in this way. Therefore, so-called extended business models are referred to in this respect, which combine functions at company level with interfaces at the level of the whole or part of the rail transport chain. The scope depends on the location, tasks and role of the operating entity on the railway market. Such an interdisciplinary approach to designing digital business models in the context of their cooperation with railway systems creates a new paradigm of cooperation in the railway market within the boundaries of the railway system focusing on a new paradigm of safety management mechanisms using the digitalization of railways and the innovative digital business models of railway enterprises, covering the entire range of railway services including auxiliary processes such as the control and management of railway traffic, maintenance, along with the diagnostics of railway infrastructure and railway vehicles. In general, the digital business models can support various processes pertaining to railway safety—they may include support for classical railway traffic control systems including remote control systems and devices, systems for recording data from the operation process such as signaling/train dispatching assistance systems, systems providing operational communication between railway stations, safe train operation control systems, automatic warning systems (AWS), automatic train protection systems (ATP), automatic train control systems (ATC), automatic train operation systems (ATO), fire safety and access protection systems and devices, electrical safety systems and devices, systems and devices to detect failures and malfunctions of rolling stock, panic prevention and rescue support systems and devices, and others (Pawlik 2019, pp. 53–96). The above-mentioned areas of the digitalization of railway traffic control, operation and support processes are often the basis for creating the innovative business models of enterprises which deliver these solutions to the railway market. These business models not only attempt to deliver value within the identified areas, but above all they are focused on innovative ways of delivering the expected value resulting from the proposed solution, including the service, supervision and support of processes, the coordination of solutions with other solutions, data collection and their effective use and other attributes dedicated to the goal.

3.4 Social Business Models in Rail Transport and Digital Transformation

With a clear message and a clear interpretation, it is worth noting that the rail transport sector operates in the so-called interactive socio-technical system. This means that inter-sectoral, inter-branch and intersubjective relationships in social and technological dimensions are crucial. The interpretation of that reasoning at the level of the railway enterprise is the business model used. As safety can certainly be defined as a social factor, the determinant of the understanding of the business model of such an enterprise is social criteria. They are built into the business model, which is a platform for the construction and implications of the social model of business in rail transport. The question arises as to what the social business model in rail transport is. It is a source of multidimensional value proposition for rail transport users; it is a new creator of numerous value propositions for customers in the sector and a tool for generating sustainable results in terms of reasonable profit. It is considered in terms of justified costs and expected results in a regulated sector such as the rail transport sector. The social business model of a railway enterprise is therefore an ontological entity which is relevant to its structure and function, which ensures the achievement of social, economic and technological results by means of a coherent configuration of these factors built up from a component canvas of this model, which ensures the fulfilment of its key objective, namely to achieve the economic objectives related to operating railway traffic or train and maneuvering operation in a safe manner. The distinguishing feature of the social business model of a railway enterprise from the new development perspective is its ability to create a dynamic digital transformation.

3.5 Sustainable Business Models in Rail Transport and Safety Criteria

Dynamic balance in rail transport is an important attribute of its existence in the organizational, management, economic, social and technological dimensions. The pursuit of a sustainable transport system throughout the supply chain is a fundamental need embedded in a modern understanding of the flow of goods and services both locally and globally. From this perspective, it is important to develop assumptions for constructing a business model for a railway undertaking and an infrastructure manager, who use the principles of constructive confrontation within the framework of the applied dynamic balance, i.e. the sustainable business model. The use of sustainable business models in rail transport has a clear message. The rail transport sector is regulated and is therefore subject to a number of legal regulations that determine the logic of its existence. Thus, in economic terms, railway enterprises are mainly focused on generating expected revenues and justifiable costs, the aim of which is to achieve a balanced profit. In the context of safety conditions, this is extremely understandable, as the biggest challenge for rail operators is to combine business objectives with safety objectives; inter alia, this exhausts the mechanisms of constructing sustainable business models. In this sense, the achievement of business objectives in a safe manner by the railway sector is strictly justified and in line with the assumptions for the high reliability organization. It is important to define the rules for conducting business activity by railway enterprises against the background of crucial railway traffic safety challenges. Hence, constructive confrontation between the following is important:

- risks arising from the operation of railway traffic and market opportunities,
- safety objectives and business objectives,
- operational safety-related processes and business processes,
- safety performance and business performance.

A subjective perspective of the understanding and application of the sustainable business model by railway enterprises is also necessary. This concerns building a dynamic balance in creating incentives and making a significant contribution to the development and growth of railway enterprises from their network of stakeholders. Therefore, constructive confrontation in management processes is needed between shareholders, employees, suppliers, customers, legal bodies, influential institutions, co-operators and other stakeholders around the rail transport sector.

3.6 Digital Business Models in Rail Transport in the Context of New Economies

Rail transport is currently being shaped by digital solutions. On the other hand, the growth dynamics of market changes in the rail transport sector are determined by the degree to which the potential of new economies is utilized, which creates a multidimensional area of change both at the level of economic sectors and specific companies operating in a difficult market. The digitally changing world opens new perspectives for rail transport. Linking rail transport and its mechanisms to, inter alia, the sharing economy, the circular economy, the Big Data economy, cloud computing, artificial intelligence, and the social media economy generates new added value, both technologically and socially, for a wide range of stakeholders in the sector. It should be noted that such an approach is a huge challenge for rail transport creators. In the case of the digital business model and the sharing economy relationship, a huge revolution can be seen in complex logistics chains, where the mechanisms of renting, lending, and sharing not only vehicles of different types, but also a wide range of infrastructure resources such as warehouses and workshops, are already standard. This not only ensures maximum return on investment through better use, but also an additional source of revenue, leading to the creation of new resources while at the same time making better use of current resources. As regards the digital business model and the circular economy relationship, it mainly concerns the creation of green logistics chains in rail transport and the use of green modes of rail transport with the implementation of the principles of e-mobility or intelligent mobility, including intelligent transport systems (ITS). It then becomes logical to use the so-called mechanisms based on the construction of integrated, sustainable transport systems, which are at the same time a crucial element in the construction of smart cities. It is worth noting that achieving environmental results in rail transport also involves the use of renewable or alternative energy sources in their business models. This is demonstrated by, inter alia, the use of hydrogen-powered units. With regard to the relationship between the digital business model and Big Data analytics, the following concepts are dealt with: the connected commuter, Mobility as a Service (MaasS), Predictive Maintenance as a Service (PMaaS), the automation and interoperability of traffic control systems (GoA4) and the Internet of Trains. In the context of digital business models and artificial intelligence logic, this is, inter alia, related to the use of autonomous vehicles and the introduction of precise train punctuality management systems. It is worth noting that extremely strict safety requirements apply in rail transport. Marked reluctance arises regarding the replacement of a human who makes decisions in real-life critical situations with a machine.

3.7 Conclusion

The digital economy is forcing a new perspective on the functioning of the rail transport sector. Railway enterprises are shaping their business models through digital transformation. At the same time, railway systems are being digitalized, in many cases using the functionalities of the digital business models of railway enterprises. In numerous cases, the boundary is blurred between what comes from the digital business model of railway undertakings and what constitutes a railway system, understood as the rail network distinguished by the functional and technical characteristics and railway vehicles intended for traffic on that network. In this situation, it is not necessary to ontologically separate both these spaces. There is also a strong relationship between the assumptions of the socio-technical system, which is the railway system (the organized structure of employees coupled with the technical system) and the social aspects of the business models of railway enterprises. The common connecting denominator, which is a social factor, is rail traffic safety, where safety determinants stimulate the shape and form of digital solutions. From a theoretical point of view, these assumptions come from the knowledge and scientific achievements of cybernetics, which is the science of control derived from the systems theory. The combined process of designing digital business models and the digital transformation of the railway system influence the social impact of the entire supply chain, making it sustainable. This sustainability involves, inter alia, building a knowledge and information exchange platform for the efficiency of the entire transport system. From this perspective, the relationship between social and technical factors, which has a strong economic impact on the operational and investment activities of railway enterprises, is crucial.

References

Arbib MA (2018) From cybernetics to brain theory, and more: a memoire. Cogn Syst Res 50:83–145. https://doi.org/10.1016/j.cogsys.2018.04.001

Asaro P (2010) Whatever happened to cybernetics/Geist in der Maschine. Verlag Turia, Wien

Chen L, Huang L, Li C, Wu X (2017) Self-adaptive architecture evolution with model checking: a software cybernetics approach. J Syst Softw 124:228–246. https://doi.org/10.1016/j.jss.2016.03.010

Chepin E (2021) Robotics: from first-order cybernetics to third-order cybernetics. Procedia Comput Sci 190:130–136. https://doi.org/10.1016/j.procs.2021.06.016

Iwachnienko AG (1962) Cybernetyka Techniczna. PWN, Warszawa

Lepsky VE (2015) Economic cybernetics of self-developing environments (third-order cybernetics). Manag Sci (Financial University Under the Government of the Russian Federation) 4:22–33

Levin-Rozalis M (2010) Cybernetics: a possible solution for the "knowledge gap" between "external" and "internal" in evaluation processes. Eval Program Plann 33:333–342. https://doi.org/10.1016/j.evalprogplan.2009.12.007

Liu L, Zhou Q, Liu J, Cao Z (2017) Requirements cybernetics: elicitation based on user behavioral data. J Syst Softw 124:187–194. https://doi.org/10.1016/j.jss.2015.12.030

Lopez J (2014) Digital business is everyone's business. Forbes. Available via Forbes. http://onforb.es/1uBkChg. Accessed 09 November 2021

Mazur M (1966) Cybernetyczna teoria układów samodzielnych. PWN, Warszawa

Mazur M (1976) Cybernetyka i charakter, 1st ed. PIW, Warszawa

Mulder M, Pool DM, Abbink DA, Boer ER, van Paassen MM (2016) Fundamental issues in manual control cybernetics. IFAC-PapersOnLine 49(19):1–6. https://doi.org/10.1016/j.ifacol.2016.10.429

Mynarski S (1981) Elementy teorii systemów i cybernetyki. PWN, Warszawa

Novikov DA (2016) Cybernetics: from past to future. Springer, Heidelberg

Osterwalder A, Pigneur Y (2010) Business model generation. A handbook for visionaries, game changers, and challengers. Wiley, Hoboken, NJ

Pawlik M (2019) Referencyjny model funkcjonalny wspierania bezpieczeństwa i ochrony transportu kolejowego przez systemy z transmisją danych. IK, Warszawa

Schuck TM (2021) Cybernetics, complexity, and the challenges to the realization of the internet-of-things. Procedia Comput Sci 185:45–54. https://doi.org/10.1016/j.procs.2021.05.040

Wei YP, Li XM (2019) Forecast of railway passenger transport turnover based on GM(1,1) model. In: ICEMME 2019, pp 347–350. https://doi.org/10.1109/ICEMME49371.2019.00075

Wiener N (1948) Cybernetics: or control and communication in the animal and the machine. Hermann & Cie, Paris and MIT Press, Cambridge, MA. ISBN 978-0-262-73009-9; 2nd revised ed. 1961

Yang H, Chen F, Aliyu S (2017) Modern software cybernetics: new trends. J Syst Softw 124:169–186. https://doi.org/10.1016/j.jss.2016.08.095

Chapter 4
Safety Management in Rail Transport: Theoretical Assumptions and Practical Implications

Adam Jabłoński

Abstract *Purpose*: The aim of this chapter is to present the key assumptions for understanding rail transport safety management.

Design/Methodology/Approach: A critical literature review and theoretical and practical reflections aimed at explaining the complexity of technological rail transport management, have been adopted as the methodology for this chapter.

Findings: This chapter presents the interdisciplinary nature of rail transport safety issues.

Research limitations/Implications: The limitations of the analytical and evaluation study stem from the difficulties in identifying those issues which are of the utmost importance in fully explaining the role of safety management against the background of specialist issues and contextual safety determinants of railway systems.

Practical implications: The practical nature of the issues discussed has been demonstrated in the resultant nature of the proposed methods and approaches to attempts to clarify the selected aspects of shaping approaches to rail transport safety.

Originality/Value: The originality of the analytical and research intention is the comprehensive identification of various concepts such as the role of railway market operators, the specificity of the functioning of the railway system, the value chain in rail transport, the safe operation of railway vehicles and the place and role of safety management systems.

4.1 Introduction

The mechanisms of effective and efficient rail transport safety management are based today on two crucial, technological and digital forces. Technological force is determined by various technologies and their configurations used in the processes of railway traffic management and infrastructure management within a complex technical system. By contrast, the digital force complements the technological force with digital solutions which interact with technology. Both of these forces in dynamic

A. Jabłoński and M. Jabłoński, *Digital Safety in Railway Transport—Aspects of Management and Technology*, Springer Series in Reliability Engineering,
https://doi.org/10.1007/978-3-030-96133-6_4

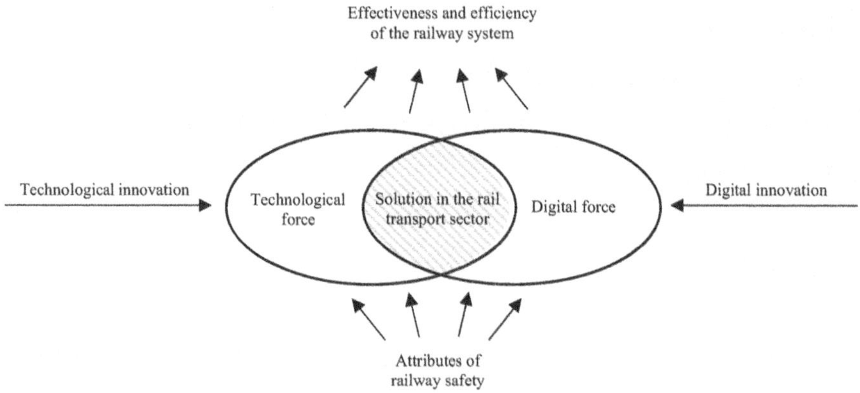

Fig. 4.1 The ability of technological and digital force to shape rail transport safety (*Source* Own study)

terms are, in a sense, drivers for the development of innovation in the rail transport sector towards safe, efficient and effective railways. Current rail transport solutions are the result of this perspective. Figure 4.1 shows the relationship between technological and digital force in shaping railway safety.

It should be borne in mind that the outcome of effective rail transport safety management is a state in which there are no risks, in this case physical and digital risks, often caused by different factors. Therefore, it is necessary to consistently implement safety objectives for many entities in the rail transport sector, encompassing both technological and digital determinants of shaping rail transport safety.

4.2 The Place and Role of Rail Market Entities in Terms of the Allocation of Responsibility for Safety

The rail transport market is regulated by 'unbundling', which consists of separating the activities of the railway infrastructure manager from the tasks of the railway undertaking. Regulations in this regard vary slightly in different European Union countries. In general, however, it is crucial that responsibility for making railway lines available on a competitive basis is entrusted to infrastructure managers, whose task is to manage railway traffic and maintain railway infrastructure in a safe manner. In turn, the railway undertaking should focus exclusively on the carriage of people or goods and comply with the rules governing the management of railway traffic on railway stations and lines. In addition to infrastructure managers and railway undertakings, the railway market identifies the needs of the functioning of entities such as the Entity in Charge of Maintenance (ECM) to maintain railway vehicles in their life cycle, taking into account parameters such as safety and operational readiness. The railway market is regulated and is supervised by national organizations which perform the tasks of the National Safety Authority (NSA). They conduct activities in accordance with European Union railway regulations published by the European Union Agency for Railways. The Agency presents prospects for the development of the rail sector in Europe, discusses them with representatives of the Member States of the European Union, and sets standards of conduct. It also performs tasks related to the implementation of innovative solutions throughout the European Union on the basis of established technical, operational and organizational standards. A consensus approach based on dialogue, consultation and agreements is preferred in the development of final solutions relevant to the specificity of the railway market in the European Union. The system of responsibility is abundantly clear and results from a number of legal regulations which, both in terms of the scope of tasks and the manner of implementation, determine the conditions for the operation of the rail transport sector. In addition to the above-mentioned core railway market participants, entities such as rolling stock repair workshops, manufacturers, notification bodies (NOBOs) and safety integration assessment bodies (ASBOs) are active in operational terms. Each of these activities is defined and described in detail in numerous regulations. Such a ranking of responsibility in the operation of the railway system is characteristic of sectors whose activities are complex. The rail transport sector is such a system, being complex in legal, organizational and technical terms. It should also be noted that the rail market in Europe is highly liberalized. Many infrastructure managers operate on the market who are embedded in a defined geographical area, as do railway undertakings who do not have such a restriction. As part of an approach based on ensuring unrestricted access to railway infrastructure, legal regulations governing the railway market allow railway entities to operate throughout the Community, while respecting the relevant railway market safety, interoperability and competitiveness requirements. The crucial role of railway sidings, the task of which is to ensure safe and efficient transshipment and maneuvering processes, is also worth mentioning. The potential and quantity of railway sidings prove the attractiveness of the railway system. The more transshipment sites fulfill the highest standards, the better the railway system is. The subjective perspective which defines the tasks of railway market participants must be consistent with the characteristics and conditions of the infrastructure and the geographical location. This applies to parameters such as the density of the railway network, the average operating speeds, the status of railway lines and the technical condition. From this perspective, entities operating on the railway market should comply with the technical and legal requirements which enable them to conduct their business activities safely and in accordance with the law. The safe operation of the railway system will therefore depend on the responsibility of individual railway market operators and the individual organizational culture within which safety culture is shaped. The railway system as a whole will be safe when all rail market participants ensure within their scope of activity that legal requirements are fulfilled at all times and assume responsibility for rail traffic safety. Therefore, the procedure for the certification of the safety management systems of railway infrastructure managers and railway undertakings, who are required to operate on the basis of specific safety documents issued, is applied. The evidence of safety for infrastructure managers is safety authorization, a safety certificate for railway operators, and for railway siding users a safety certificate for railway sidings. In addition, ECMs are also certified. They are required to obtain a certificate confirming the functioning of the maintenance management system for railway vehicles in accordance with the regulations resulting from the safety and interoperability directives. All these circumstances and requirements mean that the railway system in the context of its operators is strongly regulated by law and the roles of its users in business terms are divided in detail. Safety is also shared at that time, since it is the responsibility of each operator in its area of operation to ensure the safe operation of the railway system. Therefore, when an undesirable railway incident occurs, the causes of the incident should be carefully investigated and those found to be guilty or complicit in the incident should be identified in this respect. Such a division of responsibilities seems appropriate in terms of the conditions of competitiveness, access to the railway market and safety criteria.

4.3 The Railway System and Its Technical, Operational and Organizational Determinants

The traditional safety analysis of railway systems focused on the usefulness of individual systems resulting in individual assessments. Today, systems usually do not work in isolation, if they did at all (Barnatt and Jack 2018). The rail transport sector operates within a dynamic, interactive and complex technical system. The dynamics of this system are mainly linked to the fact that the system changes in the designated function of time and space. It is possible to clearly define the logical transition from a subsystem to another subsystem within a complex system in a spatial arrangement. These assumptions determine the specific technical, operational and organizational conditions of the system. The complexity of this system is determined, inter alia, by the interpenetration of these determinants. As this system is composed of multiple components aggregated together, the crucial factor in meeting its assumptions is the coherence and integrity of the system in the context of the needs of its expected functionality. In order for the railway system to enable trains to be driven within the serviced railway infrastructure, it must comply with the fundamental requirements. The crucial premise is the operation of railway traffic so as to eliminate any interference which may be systemic or incidental and result from the operation of or interference with the railway system itself. These disruptions may result in railway incidents, i.e. disasters, major accidents, accidents and incidents. The system must therefore be tight, resulting in its ability to conduct stable and sustainable rail traffic. The tightness of this system is intended primarily to guarantee railway traffic safety, i.e. a condition in which there are no risks. This is done through the management of the railway system in such a way as to achieve the safe driving of trains by the railway undertaking, the safe operation of railway traffic by the infrastructure manager and the safe operation of railway vehicles by entities in charge of maintenance (ECM). Only such symbiosis and synergy between these key players can

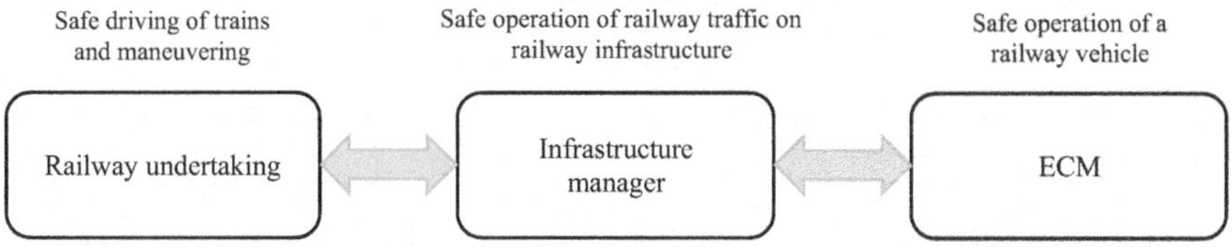

Fig. 4.2 Relationship between railway traffic participants in subjective terms (*Source* Own study)

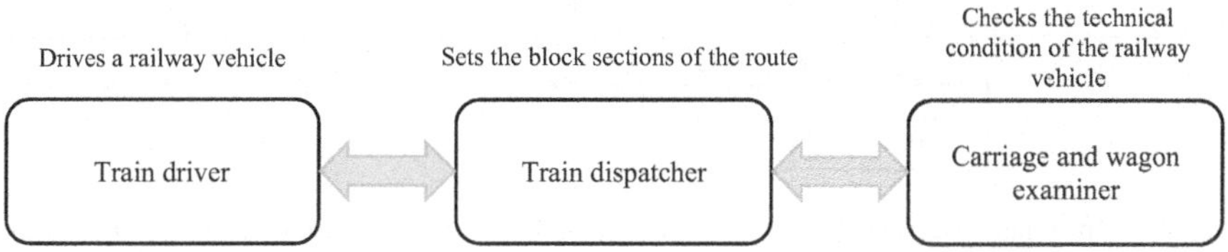

Fig. 4.3 The relationship between railway traffic participants from the personal perspective (*Source* Own study)

develop appropriate mechanisms for the operation of an efficient and effective rail transport system. Figure 4.2 shows this relationship in subjective terms.

From the personal perspective in the railway system, a relationship between the train driver, the train dispatcher and the carriage and wagon examiner occurs (Fig. 4.3). These positions perform railway traffic safety–critical tasks, and they must be under special supervision.

In terms of ensuring the operational capacity of the rail transport system, the objective perspective appears, in addition to the subjective perspective, where railway traffic participants are involved. It can be determined by the relationship between railway traffic control, railway infrastructure and energy, where railway traffic is conducted, railway infrastructure is maintained and railway vehicles are maintained in an acceptable state. The so-called interoperability of railways is then created (Fig. 4.4).

The result of this solution is that its technical compatibility and safe integration are ensured within the operating railway system.

4.4 Technological Management in Terms of Improving the Performance of the Technical Railway System and Safety Criteria

The rail transport sector is a high-tech sector. This is mainly due to the complex and complicated structure of the transport system. From this perspective, technology and its application in rail transport are the foundations of its operation and specific

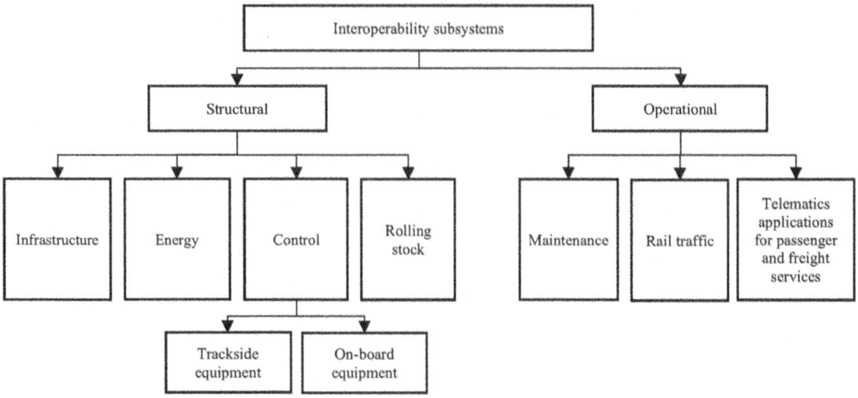

Fig. 4.4 The relationship between railway traffic participants in objective terms (*Source* Own study)

management mechanisms based on technological management processes. Technological management then becomes the determinant of both ensuring the continuity of the functioning of the railway system and the performance of its essential functions and performance measured, inter alia, by the efficiency of that system. The key parameter measured and expected in this case is the technological efficiency of the railway technical system as the result of a complex measurement system serving that technical system. In order to be able to manage the technologies in an organization, it is necessary, first of all, to assess the degree of its technological advancement compared to global solutions and to assess the achievements of its competitors in this regard. In this context, various methods dedicated to identifying the level of advancement of research and development projects aimed at developing innovative technical solutions can be used. In rail transport, the Technology Readiness Levels (TRL) methodology (Mankins 1995) and Engineering Manufacturing Readiness Levels (EMRL) can be used to conduct a full analysis of possible applications that improve, inter alia, the safety of technological solutions (Defence Research and Development Canada 2006). In order for railway enterprises to develop technologically, it is essential to conduct the so-called preliminary technological audit. A technological audit can be understood, inter alia, as a method for assessing the technological capability, procedures and needs of enterprises. Moreover, the strengths and weaknesses of the organization are identified through the description and overall assessment of the company's core/basic know-how (marketing, management, finance, human resources, etc.). This is a process of analysis that leads to the development of specific proposals (an action plan). The technological audit is conducted by external consultants in close cooperation with the company's management and staff (Kelessidis 2001). The final result of the technological audit is the assessment of the technological state and capacity of the organization, as well as the identification of its technological needs. It also indicates the areas for improvement and identifies strengths and weaknesses as well as potential opportunities and threats of technologies along with their forecasting and positioning in relation to the level of technology development of competing companies. This discussion shows that each railway enterprise, as a high reliability organization, must systematically conduct technological audits in order to be able to determine its development and technological needs in a logical and systematic manner. First of all, it is worth paying attention to the relationship between the two key parameters applicable to railway enterprises, i.e. technological advancement and safety level. The assessment of technological readiness (U.S. Department of Defense 2016) provides managers with reliable information on the state of research and development.

TRL level 1: refers to the fundamental principles of the new technology. This is the lowest level, which means starting research into the basic characteristics of the technology.

TRL level 2: the concept of technology has been formulated, the possible use of new technology has been described, but there is no experience and thorough analysis to confirm the correctness of the concept.

TRL level 3: analytical and laboratory tests for the correctness of the concept have been conducted. At this level, the research and development phase begins, which includes the development and description of the technology, and laboratory tests which aim to verify existing research.

TRL level 4: the technology concept and its parameters in laboratory conditions have been verified, and the basic components of the technology have been integrated. A general mapping of the final product/technology has been obtained under laboratory conditions.

TRL level 5: the concept of technology and its parameters have been tested in an environment similar to the actual one. The basic components of the new technology are integrated into the actual components of the system. The technology may also be tested under simulated operating conditions.

TRL level 6: the operation of the prototype technology has been presented in conditions similar to actual ones. Research may be conducted under laboratory conditions that reflect the actual conditions to a high degree.

TRL level 7: the operation of the product/technology prototype in operational conditions has been presented. This level identifies significant progress in technology maturity and requires substantial financial resources in a large number of areas.

TRL level 8: the technology has been assessed, and the fulfillment of design assumptions has been confirmed—the technology has reached the expected level and can be used in the expected conditions.

TRL level 9: the new technology is already in its final form and can be implemented in target processes; it has been tested in operational conditions with a positive result (production).

The Safety Integrity Level (SIL) and the Performance Level (PL) are parameters which determine the safety performance of control systems related to the safety of

Table 4.1 SIL levels with a defined measure of the number of failures/errors defined by tolerable hazard rate (THR)

SIL level	THR value
Continuous operation	
4	$\geq 10^{-9}$ and $< 10^{-8}$
3	$\geq 10^{-8}$ and $< 10^{-7}$
2	$\geq 10^{-7}$ and $< 10^{-6}$
1	$\geq 10^{-6}$ and $< 10^{-5}$
Work on demand	
4	$\geq 10^{-5}$ and $< 10^{-4}$
3	$\geq 10^{-4}$ and $< 10^{-3}$
2	$\geq 10^{-3}$ and $< 10^{-2}$
1	$\geq 10^{-2}$ and $< 10^{-1}$

Source IEC 61508-1:2010 Functional safety of electrical/electronic/programmable electronic safety-related systems—Part 1

machinery. SIL levels are set for electrical and electronic (including programmable) safety systems, while PL levels are also used for other pneumatic, hydraulic, mechanical and less complex electrical systems. Two key standards are used to directly assess the level of safety: EN 61511 series—Functional safety—Safety instrumented systems for the process industry, and EN 62061—Safety of machinery—Functional safety of safety-related electrical, electronic and programmable electronic control systems. It should be noted that SIL levels are determined in continuous and on-demand modes. Both of these options are present in rail transport processes. SIL levels are used in the design of devices that must fulfill the safety requirements are shown in Table 4.1.

Standards related to specific domains (railway traffic control, production process control) specify where and what level should be adopted, e.g. railway traffic control systems whose reliability parameters are defined in the EN 50129 standard should comply with SIL 4. Any safety-related activity focuses on ensuring safety at a specific stage of system design and/or by considering specific hazards in the system itself. Hazard identification concerns the development of failure scenarios; risk assessment is intended to predict the occurrence of failure scenarios; while verification activities are used to reduce the frequency of dangerous scenarios (Bolbot et al. 2019). Therefore, it is interesting to conduct a multi-criteria analysis comparing the Technology Readiness Level (TRL) of technological advancement and Safety Integrity Level (SIL) in pairs. The concept of safe integrity becomes a connecting factor in this area. Safe integrity combines two fundamental concepts: integrity in the event of system failures and integrity in the event of random failures. The former represents failures in specification, identifiability, design, production and maintenance caused by a failure or incorrect human process, while the latter appears randomly due to the failure of mechanical-electronic devices, the aging process or wear and tear. The key to this particularly important integration of TRL and SIL is to combine the assumptions of the systems theory with a complete and precise definition of the architecture of a given system together with its adopted design mechanisms. The above relationships can be supported for rail transport by comparison with four grades of automation (GoA) in rail transport:

GoA 0: Manual operation with no automatic train protection,
GoA 1: Manual operation with automatic train protection (such as PTC),
GoA 2: Semi-automatic train operation (STO),
GoA 3: Driverless train operation (DTO),
GoA 4: Unattended train operation (UTO),

described in the IEC 62290 international standard, where the highest (fourth) grade refers to a system in which vehicles are operated completely without operational personnel on board. This means that in many cases the goal should be to develop final solutions for railways, where their parameterization would include: TRL 9, SIL 4, and GoA 4.

4.5 Information Management in the Rail Transport Chain and Railway Traffic Safety

The flow of information in complex socio-technical systems, and particularly the railway system, is essential to ensure the expected level of rail traffic safety. Information management can be understood as a set of activities involving people and technical systems to effectively transfer information from one place to another. The word 'management' in this context is designed to integrate both the human factor and its verbal reactions, as well as the technical system notifications and responses within the human-technical system interface. Information management in the rail transport chain plays a crucial role in the business continuity of the railway system. Communication within the railway system primarily concerns technical and operational processes and requires the flow of specific information that allows for the continuation or interruption of key processes pertaining to railway traffic or to processes supporting such railway traffic. Technical systems in rail transport have different levels of complexity, come from different periods, rely on various technical platforms and differ in terms of their specific performance and functionality. In many cases, information systems are based on different ways of encoding data, other ways of processing them and different information flow functions. The multiplicity of traffic control systems, in particular, results in the asymmetry of the information flow—that is to say, the uneven flow of information between traffic control systems which interferes with the symmetry of the information being processed. One system requires less basic information while the other system interacts with it with a much greater need to process information. This situation means that one operator has to process much more information, which increases the chances of making errors. It is therefore reasonable to integrate railway traffic control systems in such a way that the level of symmetry of data processing is similar and preferably identical. For interfaces between various devices, this negative asymmetry of data processing will always occur. In the area of the specificity of rail transport in the context of information management, attention should also be paid to parallel processes, which also generates the need for a multi-directional flow of information. This is conducive to the risk of losing information, not using it properly or using it at the wrong time. Difficulties in managing information can then arise from the dispersion of data across the railway system in terms of three key layers: the information layer in terms of the occupancy of the track system, the information layer in the field of railway traffic control and the information layer in terms of the planned railway traffic flows in the logistical sense. These three core layers of information flow can affect both the expected level of railway safety and the efficiency of the transport system. The effective exchange of information in this respect is essential to ensure the expected level of railway traffic safety. Any error in the flow of information can be the cause of negative railway incidents. The railway system is mainly based on the exchange of information due to its specificity. Activities in the railway system are always preceded by the exchange of information, not only between people but also between artificial creations such as individual rail transport subsystems. In modern railways, parallel communication between people and the structural and functional subsystems of rail transport occurs. Interface management takes place at the level of mechanical, electrical, electronic and programmable systems. This is then based on the concept of functional safety—ensuring the proper functioning of the safety-related control group systems. The safety circuits of machinery, processes and installations should always be linked to evidence of the ability to perform functions under foreseeable conditions. Technical linkage systems must be supported by an appropriate flow of information in a dynamically operated technical system. Therefore, safety issues should be supported by various information management solutions.

4.6 The Assumptions of the Functioning of Railway Traffic Control and Management Systems in Terms of Safety

Railway traffic management systems are one of the most crucial aspects of the digitalization of rail transport systems. The approach to railway traffic control has evolved since the beginnings of railways. The first railway traffic protection systems were primitive and consisted of protecting the main tracks from train overruns from the siding and determining the occupancy status of the tracks. These functions are still maintained but the functional range of railway traffic management systems is growing as technology develops. The capacity of railway traffic control devices and systems is also evolving. First of all, the goal was to secure railway traffic—that is, generally speaking, to provide protection against danger. It is for this purpose, for example, that derailers are still used on the side tracks. Another word which relates to this issue is control, which can generally be understood as directing the work of a device using appropriate devices. Railway traffic management, on the other hand, has a much wider connotation. Traffic management is a set of actions taken to make the best possible use of transport infrastructure to ensure the safe and efficient movement of people and goods. The evolution of the approach in this respect therefore involves the shift from railway traffic protection through railway traffic control to railway traffic management. The same is true with the use of railway traffic control devices. At the beginning, they were mechanical devices, including those based on keys to confirm train runs; then mechanically centralized; at the stage of further development these were the relay equipment; and finally, digital devices were created. The configuration of these devices has always aimed to create a coherent system, and the primary objective of such systems has always been to achieve the expected level of railway safety. These systems therefore focus on the train operation in terms of safety. To this end, they are divided into Class A systems and Class B systems. Class B Control Command and Signaling (CCS) systems are national systems which existed before the Interoperability and Safety Directives entered into force. This class includes the Automatic Train Braking System. The Class A system is the European Rail Traffic Management System (ERTMS), which was introduced in the 1990s. Its aim was to ensure railway safety, support interoperability between national rail networks and cross-border rail transport, reduce the cost of purchasing and maintaining signaling systems and increase the infrastructure capacity and reliability of rail transport systems. It is estimated that in Europe the fully deployed ERTMS would increase capacity by up to 30% using existing infrastructure. The deployment of ERTMS has to include the withdrawal of Class B systems from ETCS trackside elements and the extensive modernization of rolling stock by equipping it with interoperable subsystems with ERTMS on-board equipment. The establishment of a railway traffic control system framework to ensure the involvement of railways in the digital transformation which ERTMS is based on should enable easier and cheaper updates and modernization by means of greater standardization and modularization, the introduction of plug-and-play principles and the harmonization of data modelling. The digitalization of railway traffic control solutions is the most important aspect of the digital transformation of railways. Such trackside and on-board control command and signaling subsystem solutions also generate digital transformation needs in other structural and functional subsystems such as infrastructure, energy, rolling stock, railway traffic, and telematics. In this way, modern railways are moving dynamically towards full digitalization and improved railway traffic safety. It should be noted, however, that a large number of safety studies focus exclusively on control systems, almost completely excluding other systems used to ensure rail transport safety, while the number and diversity of such systems is clearly increasing (Pawlik 2019, p. 14). Thus, safety management should be seen in a holistic manner as using various solutions in both the safety and security of railway systems, the functions and performances of which enable the desired level of safety to be achieved in the technical, operational and organizational layers. All solutions should be included in the full range of the operation chain of the railway system, where railway traffic control plays a crucial role.

4.7 Safe Operation of Railway Vehicles—Key Determinants

Railway vehicles are an essential element of the railway system and they form an interface with other structural and functional subsystems. In the context of power supply, they form the interface with the energy subsystem in terms of the contact wire and the pantograph. A railway vehicle runs on tracks by means of the rail head and the wheel set interface. Modern digital railway traffic control systems interact in terms of the on-board devices of the vehicle and track-side command-control and signaling systems within the ERTMS solution. In addition, GSM-R-based communication systems require the installation of such devices on vehicles. The most modern railway vehicles can communicate with each other. In this way, it is in fact possible to ensure the safe operation of railway traffic only with the participation of vehicles (enabled by ETCS level III). The safe operation of railway vehicles is therefore the foundation for ensuring the expected level of railway safety. In this respect, the technical compatibility of the vehicle with the network, the technical compatibility within the vehicle and the safe integration of the railway vehicle into the rail network must be ensured. Safe operation includes in particular the vehicle driving process but also the preventive maintenance process, which ensures the expected level of operational readiness and safety. Both of these parameters must be monitored in terms of the efficiency of railway vehicles throughout the railway system. As regards ensuring the safe operation of railway vehicles, attention should be paid to their wide diversity in the railway system. They are divided into freight traffic vehicles and passenger traffic vehicles. Naturally, each of these types has a different specificity affecting railway traffic safety. Freight transport, due to its specific characteristics, runs at lower speeds than passenger transport, which, when operating railway traffic connecting freight and passenger transport, generates a lot of risks in terms of safety; however, it also significantly affects the efficiency of the use of railway lines. This is related to the large number of disruptions that may arise because freight and passenger traffic are mixed. Such problems do not occur on railway lines intended solely for freight or passenger traffic. The implementation and use of railway vehicle maintenance management systems by entities in charge of maintenance (ECM) is worth noting. The regulations required by law in this respect oblige railway undertakings and infrastructure managers to appoint a specialized external unit or to independently maintain railway vehicles in accordance with legal provisions. The requirements in this regard are extremely strict due to the crucial place and role of railway vehicles in creating hazards in rail traffic. The safe operation of railway vehicles is multidimensional and requires a broad view. The digitalization of these processes is reflected in the development of innovative applications for monitoring the location of railway vehicles, the status of their maintenance processes, detection of rolling stock defects and failures, the use of digital solutions for the predictive maintenance of railway vehicles and support for the operation of the technical facilities in terms of logistics processes. There is plenty of room for digital transformation in this area.

4.8 The Place and Role of Railway Traffic Safety Management Systems

Safety management systems are not a revolutionary concept or a means that have emerged spontaneously. Their appearance is closely related to the evolution of thinking about safety (Bieder 2021). Safety management systems are crucial to ensure the expected level of railway system safety. Their formal character allows for the identification of all the technical, operational and organizational links intended to cover all aspects of the activities of entities which operate within the railway system. These systems are a requirement for railway undertakings and infrastructure managers, all of whom are required to comply with legal regulations which sanction the need to implement and maintain safety management systems. They combine all the processes which the above-mentioned railway market operators are responsible for. Their premise is to identify the hazards and risks that should be mitigated by different control measures. A systemic approach to operational risk management is a crucial element and distinguishing feature of this approach to railway traffic safety. The legal regulations pertaining to safety requirements in terms of management systems are based on Directive 2016/798, which explicitly states that railway undertakings should conduct their business in a safe manner. The mission of railway enterprises should be a strong focus on ensuring railway traffic safety while ensuring the ongoing economic viability of the enterprise. Profit should not be a priority; in this sector a high level of safety must first be ensured, and only then should profits be pursued. This should be done by identifying all key processes, the main ones being operational, management and support included in the scope of responsibility and decision making in railway enterprises. The transport process of railway undertakings and the process of maintaining and making railway lines available supported by safe railway traffic management solutions should be covered by a bundle of supporting processes that will ensure cause-and-effect supervision of operational processes. The interrelationships between processes should be described in the form of processes indicating milestones and performance indicators. Process supervision should be based on appointing process owners and process operators. The identification of actions in the processes should ensure that all safety requirements are fulfilled by railway enterprises. Process maps should indicate which process configuration the railway enterprise has adopted in order to conduct its activity in a safe manner. Therefore, safety in terms of safety management systems not only addresses technical issues but refers in particular to a holistic approach. From this perspective, errors, hazards and risks may arise in normal activity where negative events such as accidents and disasters can be caused by bad processes, with the poor allocation of responsibilities, supervision and training. The maintenance processes of crucial railway infrastructure equipment, as well as railway vehicles, may also be incorrect. It is therefore appropriate to implement organizational solutions used to structure and standardize processes in order to ensure that all activities of railway undertakings and infrastructure managers do not create dangers for railway users, passengers, railway staff and entities which interact with the railway system such as the residents of adjacent areas and other stakeholders. Risk allocation is related to the identification of different areas of activity of railway enterprises. The primary areas include operational risks pertaining to the transport process or the operation of railway traffic. Another area is the risks associated with the use of the services of suppliers and subcontractors, who may make technical or organizational errors. In another respect, these will be the risks pertaining to the maintenance of railway infrastructure and railway vehicles. An important area which should also be assessed is the area of participation of third parties in the railway system, including undesirable persons on railway tracks and road users improperly acting on railway crossings. The involvement of terrorists, including those attacking digital systems, will also be an important area of risk identification and management. The issue of the identification of cybersecurity risks is also part of the process supervision included in the description and functioning of safety management systems. The digital transformation of railway enterprises is an advanced process whereby companies attempt to use digital solutions in various links. This creates risks that have not occurred in organizations with a low degree of digitalization. As the railway system forces technological progress to some extent, safety management must therefore cover, to a large extent, the aspects of the risks posed by the operation of these devices and the exposure of companies to cyber-attacks. Safety management systems are essentially designed to support railway systems against incidents that could disrupt railway traffic and lead to dangerous situations. A holistic approach to management processes is highlighted in this context. Management itself affects all areas of the activity of railway enterprises. Therefore, the management of such an important issue as safety must relate both to the management of the entire company and to the aspects strictly related to safety issues. The overall management process of railway enterprises must be based on an implemented safety management system in which risk management is an essential and core element of this approach.

4.9 Conclusion

Railway traffic safety is of a socio-technical nature. On the one hand, it is built into technical system configurations, and on the other hand, it depends on the behavior of people who design and operate technical solutions. Railway traffic safety is influenced by many variables that need to be considered separately. However, the integration of all factors which affect safety is essential. By definition, the system must be united. Thus, a systemic approach in the context of railway traffic safety management is of vital importance. In terms of the digitalization processes of the railway system, all the achievements to date and the theoretical and application assumptions are sustained but also altered, according to the type and nature of the digital transformation. The former concerns the original assumptions embedded in the structure of the railway system, while the latter entails the need to implement new risk control measures, which must be derived from technical standards and innovative solutions subject to multiple forms of validation. Safety evolves and is constantly desirable, but the dimension, shape and understanding thereof undergo constant evolutionary processes, a dimension of which is risk orientation in relation to technological progress and people's increased levels of awareness. A broad approach to safety issues shapes the railway system as a holistic set of related components, which, both separately and together, must fulfill the expected railway safety requirements. This approach requires the specific supervision of safety management processes, and the safe integration of technical, operational and organizational solutions.

References

Barnatt N, Jack A (2018) Safety analysis in a modern railway setting. Saf Sci 110:177–182. https://doi.org/10.1016/j.ssci.2018.08.005

Bieder C (2021) Safety science: a situated science, An exploration through the lens of Safety Management Systems. Saf Sci 135:105063. https://doi.org/10.1016/j.ssci.2020.105063

Bolbot V, Theotokatos G, Bujorianu LM, Boulougouris E, Vassalos D (2019) Vulnerabilities and safety assurance methods in Cyber-Physical Systems: a comprehensive review. Reliab Eng Syst Saf 182:179–193. https://doi.org/10.1016/j.ress.2018.09.004

Defence Research and Development Canada (2006) A Technology Maturity Measurement System for the Department of National Defence. The TML System. Contract Report DRDC Atlantic CR 2005-279

Directive (EU) 2016/798 of The European Parliament and of The Council of 11 May 2016 on railway safety

IEC 6150801:2010 Functional safety of electrical/electronic/programmable electronic safety-related systems—Parts 1

Kelessidis V (2001) Technology audit. Report Produced for the EC Funded Project INNOREGIO: Dissemination of Innovation and Knowledge Management Techniques. Thessaloniki Technology Park

Mankins JC (1995) Technology Readiness Levels—a white paper. Advanced Concepts Office, Office of Space Access and Technology, National Aeronautics and Space Administration (NASA), Washington, DC, USA

Pawlik M (2019) Referencyjny model funkcjonalny wspierania bezpieczeństwa i ochrony transportu kolejowego przez systemy z transmisją danych. IK, Warszawa

U.S. Department of Defense (2016) Technology Readiness Assessment (TRA) guidance. U.S. Government Accountability Office

Chapter 5
Digital Transformation in Rail Transport—Key Challenges and Barriers

Marek Jabłoński

Abstract *Purpose*: The aim of this chapter is to present key assumptions to explain the determinants of digital transformation in rail transport.

Design/Methodology/Approach: A critical literature review with theoretical and practical reflections which aim to explain the complexity of digital transformation processes in the context of assessing the digital maturity of railway systems has been adopted as the methodology for this chapter.

Findings: This chapter identifies key approaches to understanding the place and role of the digital transformation and its impact on rail transport safety management.

Research limitations/implications: The limitations of the analytical and evaluation study are due to the difficulty in identifying those issues that are crucial to fully explaining the role of the digital transformation processes in the rail transport sector in the context of safety.

Practical implications: The practical nature of the content described lies in a critical analysis of the processes of the digital transformation of railways.

Originality/Value: The originality of this chapter is the integration of rail transport safety solutions from the hybrid perspective, which combines analog and digital solutions.

5.1 Introduction

In recent years the digital transformation has definitely accelerated dynamically and has become an obvious fact and factor of change in many sectors of the economy, as well as in the principles of society's existence, its perception and the development of knowledge and technology. Advanced digitalization capabilities provide businesses with new functionalities, greater reliability, greater efficiency and optimized capabilities, which exponentially increase the value delivered to customers (Annarelli et al. 2021). The digital transformation dynamically covered the scope of activity of railway enterprises, taking into account not only operational management but also strategic issues (Zanon et al. 2021). From this perspective, a new strategy for the development of rail transport is emerging in digital, technological and human terms.

A. Jabłoński and M. Jabłoński, *Digital Safety in Railway Transport—Aspects of Management and Technology*, Springer Series in Reliability Engineering,
https://doi.org/10.1007/978-3-030-96133-6_5

There is an undoubted change in the place and role of humans in technical transport systems. Therefore, the digital transformation not only affects analog, technical and technological solutions, but dynamically affects the human factor in rail transport and thus also probable and occurring human errors, replacing them with possible digital errors. We are now in a state of permanent and dynamic gamification between technologies, digital solutions and organizational and management determinants, with people as the entity determining the principles of safe railway traffic in rail transport. The winners in this game should be the passengers of safe and comfortable railways.

5.2 Determinants of Digital Transformation in Rail Transport

The digitalization of processes, which is now gaining wider acceptance on a global scale, is defined as the integration of digital technologies into process operations to increase efficiency, improve product quality and process safety: for example, through safety systems (Khan et al. 2021). Rail transport in this context has been changing dynamically recently. This is related to both the dynamic technological development based on innovation and the widely implemented digital transformation. Digital transformation and the resulting innovation of the business model have fundamentally changed consumer expectations and behavior, put pressure on traditional businesses and disrupted many markets. It is based on defining the strategic orientation of the organization's digital transformation and setting its strategic objectives (Zaoui and Souissi 2020). Digital transformation tends to lead to changes in company procedures and organizational structures (Fischer et al. 2020). This also applies to rail transport operators (Verhoef et al. 2021). The digital orientation dimension is the adopted configuration of the digital architecture. It captures the idea that generativity corresponds to the digital technology that triggers the mechanisms of change among numerous distributed and uncoordinated units (Kindermann et al. 2021). In this context, a new dimension of transport processes, the principles of rail infrastructure management and the mechanisms of the operation of railway vehicles in rail transport are emerging. Digitalization is fundamental in this triad of three crucial areas which shape the rail transport system. This materializes through the following determinants:

1. The evolution of the logic of managing the safety and maintenance of infrastructure and railway vehicles in rail transport.
2. The evolution of rail infrastructure management rules based on the digitalization of processes focused on the safe maintenance of railway infrastructure.
3. The evolution of rail traffic control rules based on the digitalization of processes in the requirements for setting the block sections of the route while managing rail traffic.
4. The evolution of rail traffic control rules based on the digitalization of processes in the use of modern, digital rail traffic control devices.

5. The evolution of railway traffic control rules based on the digitalization of processes in the use of specialist computer software for efficient and effective rail traffic control.
6. The evolution of railway traffic rules using specialist digital logistics systems configured with railway traffic control systems.

It should be borne in mind that digital technology can be either transformative or destructive depending on the perspective and, more importantly, the ability to exploit its potential (Saarikko et al. 2020). The positive dimension of the application of solutions always wins out, which undoubtedly has a strong impact on the efficiency of the rail transport system.

5.3 Directions and Areas of the Digital Transformation of Rail Transport

The crucial directions and areas of the digital transformation of rail transport can be identified by means of multidimensional analysis and development.

In subjective terms, this is the development of digital communication systems among individual rail traffic participants, railway undertakings, railway infrastructure managers, entities in charge of maintenance, and railway siding users.

In objective terms, this is the development of optimal configuration systems with interfaces between various interoperability technical specifications.

The directions of digital transformation in rail transport are related to technological progress in this sector of the economy. Core areas that can be selected from the numerous aspects of railway system operation include planning train timetables, planning the implementation of railway investments, planning the workload of the staff, and providing resources. As regards the sales of services within the railway system, this refers to the digitalization of marketing processes, the application of management systems, and the automatic collection of data on adverse events. In terms of network allocation, network service and planning operational plans, an operational clearing system, and a rail route reservation system may be distinguished. In the area of resource management, these are issues such as asset inventory, logistics systems, and an asset purchase system. Regarding railway signaling, the following areas subject to digitalization can be distinguished: remote monitoring systems, key management systems, registration of legal actions, temporary speed limitation systems, line and station locks, train control systems, automatic train protection systems, and radio transmission systems. As regards command control, the directions of development may be as follows: an automatic train control system, an automatic train supervision system, an energy traction system, and a wired and wireless transmission network. As regards supporting processes, the following directions of the development of digital transformation can be identified: an energy system, a HVAC system, a lighting system, and an operational telephony intercom. In terms of ensuring high-quality rail passenger services, the following systems can be distinguished: a passenger announcement system, a passenger information system, an escalator, lighting, HVAC systems, a passenger entertainment system, as well as an operational time distribution system. In terms of security, the following systems can be indicated: an access control system, a CCTV system, a network monitoring system (HOC), and a cybersecurity system. In the area of safety, the following systems can be indicated: fire detection, an emergency telephone alerting system, and operations safety systems. In terms of maintenance, these can be systems related to asset inventory, diagnostics, and maintenance scheduling systems. In terms of passenger service systems, the following can be distinguished: a ticketing system, resources allocating systems, documentation systems, and alert escalation and crisis management systems. In addition, as regards management, the following systems supporting a comprehensive railway and management system of the organization can be distinguished: HR systems, research and engineering systems, finance systems, administration systems, an administrative time distribution system, and administrative telephony. The application scope of these systems relates, in subjective terms, to the following entities: infrastructure managers' systems, railway undertakings' systems including on-board, track, station, and operations control center systems, as well as maintenance and security center systems. It is clear that the use of digital solutions covers a very wide range of activities. This takes the business aspects of the activity of infrastructure managers and railway undertakings into account. It also refers very widely to technological solutions in the key areas of operation of the railway system such as traffic control, supervision of train driving processes, maintenance, design, implementation of investments, determination of the phases of investment works, passenger service, freight transport and logistics. In fact, the entire rail transport industry is undergoing dynamic processes of digital transformation and automation. The role of people is decreasing in terms of operation, but it is increasing in terms of designing solutions, planning processes and optimizing these processes.

5.4 Assessment of the Digital Maturity of Railway Systems in Technical, Operational and Organizational Terms

The railway system is a dynamic, interactive technical system operating in a changing time and space. This system satisfies the criteria of SIS (safety instrumented systems), which consist of a set of hardware and software controls, which are used in particular in critical process systems, which undoubtedly include the rail transport system. The safety requirements of such a system are assigned to safety instrumented functions (SIFs) implemented as SIS with a defined safety integrity level (SIL). A safety instrumented function (SIF) is defined as a function to be implemented by SIS, the purpose of which is to automatically achieve or maintain a safe process in relation to a specific dangerous event. Safety function requirements are defined taking the results of hazard identification into account, while safety integrity requirements result from the analysis of potential hazardous events (Śliwiński et al. 2018). The digital maturity of railway systems is an important aspect of the digitalization of railway systems. Rail transport solutions are characterized by different levels of digitalization, which is also due to the division of the importance of railway lines. International railways, e.g. those included in the TEN-T network, will be fully equipped with digital equipment, while lower class railways may have a very low or almost no level of digitalization. The assessment of the digital maturity of the rail transport sector is an important topic in terms of both theory and practice. Maturity in such a context is usually used to assess the degree of implementation of development solutions where most modern solutions are based on the configuration of software and hardware creating uniform systems replacing existing analog solutions based on, for example, mechanical or relay devices. Digitalization is therefore intended to replace old solutions with new performances and functionalities. In doing so, attention should be paid to the sustainability of solutions. It should be pointed out that the disadvantage of digital solutions is their lesser durability when compared to mechanical and relay solutions. For example, the average life cycle of digital solutions is 15 years, relays 40 years and mechanical solutions as many as 100 years. This parameter is important but not sufficient. Digital solutions are based on better integration, rapid information exchange and automation, which in the modern world is significant in terms of the level of efficiency of the solutions used. Railway traffic operation time, the increased frequency of train rides, the comfort of operators, and the possibility of on-line diagnostics of railway systems allow for significant improvement in the safety and efficiency of the railway system. The digital maturity of the railway system should be continuously monitored and increased. The digitalization of railways is characterized by an exceptionally large number of advantages manifested by improved economic and non-economic indicators. The easier integration of IT solutions makes digitalization an irreversible and desirable process.

5.5 Integration in Hybrid, Analog–Digital Railway Systems

The need to integrate heterogeneous information resources is noticeable in railway systems. This results from the dynamic nature of the railway system and the origin of data from the numerous areas of operation of the railway system. Since the railway system is complex in nature, being constructed from an exceptionally large number of elements, it requires the construction of interfaces between devices in such a way as to enable the railway system to operate safely. In this sense, two essential problems overlap. On the one hand, many information resources from different media require mutual integration to ensure a coherent system of specific functionalities; on the other hand, this integration is hampered by a wide range of railway solutions, which differ in aspects such as other base platforms, different time and level of system complexity—they are based on different technical and operational assumptions, other maintenance modes, a different technical system life cycle, different safety classes, etc. Differing technical and organizational solutions produce a variety of forms of data in different notations that are not always possible to integrate. There is no major problem with exchanging information for those systems that are based on the same principles and come from the same production period. The situation is worse is when systems have a significant difference in performance and functionality and data migration is essential to ensure the continuity of railway traffic and traffic control. Data migration may then not be effective enough to provide all information as expected. It is also necessary to assist with additional regulations which describe the required rules of conduct in order to safely perform operational tasks despite limitations, barriers and difficulties. Efficient data and information management by means of information technology should be maintained. Dynamic organizational structures, such as socio-technical systems which constitute railway systems, are created by distributed, heterogeneous and very often autonomous units. They are shaped in an evolutionary manner based on technical progress. They are characterized by a high demand for data and information that should be provided in specific circumstances and at a particular time. They must therefore be managed. They are usually made up of computer networks included in operational processes dedicated to the achievement of the necessary objectives resulting from the needs arising from a given traffic situation in the railway system. Hybrid railway systems can be understood as heterogeneous interconnections of devices and equipment controlling the operation of railway infrastructure for the performance of tasks faced by railway systems in a given situation. The railway system in this sense fulfills the definition of a network, which includes a complex set of links geared to achieving specific strategic and operational objectives. Therefore, the migration of data and information requires an effective integration of the complex configuration of railway system components, which processes data and information, making the system dynamically efficient. Integration may concern the integration of processes, the IT applications used, traffic control and management systems and data and information resources. It is important to connect certain IT systems together with interfaces that need to be safe to ensure the safe movement of trains. A way of integrating data and information in a traditional form is to use data centralization in the formula of central databases that collect data from multiple sites, which does not prevent them from creating the complex configurations of the components that shape them. They may be modular. Another data integration formula can be based on virtual solutions where optimal hardware use, flexibility and low cost are just some of the advantages of server virtualization. Data migration can be understood as a set of actions to move data from one place where the data has already been used to a new location where they will be reused for the same or another purpose, often ensuring consistency of operation within the entire system where they were applied. In rail transport, data and information are very often shared, which means that they are processed over time as part of their migration. Information resource strategies may consist of a single shared database or distributed heterogeneous federated repositories. They may also take the form of a data warehouse. The integrated system will then refer to a strategic approach and the central database will be tactical. In the case of distributed resources, the integration process can take place on the basis of each one individually, integration through a shared scheme or application integration. The integration of the application will take the form of a service-oriented architecture, and integration through a shared scheme can be operationalized through directories, registers, applications, modified services and perspectives (Put 2012). This form of integration of information resources may be used within the framework of the identification and understanding of the operation of railway systems. Data migration in dynamic and complex technical and operational systems supported by organizational processes must be supported, first of all, by the appropriate selection of data processing tools and, on the other hand, forced by operational needs. However, decisions on the use of specific solutions must be strategic.

5.6 Conclusion

Digital transformation is a crucial challenge for the rail transport sector. New areas of railway activity are becoming digitalized. Digitalization is becoming a need arising from the rapid growth of digital solutions offered on the market or implemented directly by users in various forms of digital platforms or functional applications, and is becoming essential to preserve the necessary integration of processes, devices and systems. The requirements for the optimization of the rail transport supply chain and the need to improve railway traffic safety prompt managers to search for new areas of digitalization. It is therefore focused on improving the efficiency parameters of the railway system and railway traffic safety. An essential issue in the context of this challenge is the integration of different solutions based on different technologies, platforms and functionalities. The creation of new digital solutions therefore always requires an individual approach and the assessment of technical compatibility and compatibility of functions and synergies. An important parameter in this respect will also be the assessment of the digital maturity of railway systems against the background of dynamic robotization and automation processes. Different levels of digital maturity can be a barrier to rapid progress in service-oriented digitalization processes. Some legal developments such as the deployment of ERTMS and GSM-R in Europe stimulate both the improvement of these systems and the development of support systems. The legal requirement to adapt railway networks to comply with the assumptions of these systems generates rapid progress and the development of broad knowledge in this regard. The digital transformation of railways has become reality and there is no turning back, so attention should now be paid to ensuring that these solutions are characterized by the expected level of safety.

References

Annarelli A, Battistella C, Nonino F, Parida V, Pessot E (2021) Literature review on digitalization capabilities: co-citation analysis of antecedents, conceptualization and consequences. Technol Forecast Soc Chang 166:120635. https://doi.org/10.1016/j.techfore.2021.120635

Fischer M, Imgrund F, Janiesch C, Winkelmann A (2020) Strategy archetypes for digital transformation: defining meta objectives using business process management. Inf Manag 57:103262. https://doi.org/10.1016/j.im.2019.103262
Khan F, Amyotte P, Adedigba S (2021) Process safety concerns in process system digitalization. Educ Chem Eng 34:33–46. https://doi.org/10.1016/j.ece.2020.11.002
Kindermann B, Beutel S, Garcia de Lomana G, Strese S, Bendig D, Brettel M (2021) Digital orientation: conceptualization and operationalization of a new strategic orientation. Eur Manag J 39(5):645–657. https://doi.org/10.1016/j.emj.2020.10.009
Put D (2012) Integacja heterogenicznych zasobów informacyjnych w organizacjach dynamicznych. Wydawnictwo Uniwersytetu Ekonomicznego w Krakowie, Kraków
Saarikko T, Westergren UH, Blomquist T (2020) Digital transformation: five recommendations for the digitally conscious firm. Bus Horiz 63:825–839. https://doi.org/10.1016/j.bushor.2020.07.005
Śliwiński M, Piesik E, Piesik J (2018) Integrated functional safety and cyber security analysis. IFAC PapersOnLine 51(24):1263–1270. https://doi.org/10.1016/j.ifacol.2018.09.572
Verhoef PC, Broekhuizen T, Bart Y, Bhattacharya A, Dong JQ, Fabian N, Haenlein M (2021) Digital transformation: a multidisciplinary reflection and research agenda. J Bus Res 122:889–901. https://doi.org/10.1016/j.jbusres.2019.09.022
Zanon GN, Szejka AL, de Freitas Rocha Loures E (2021) Towards an integrated MCDM and BSC method to support the digital transformation strategy in railway companies. In: Newnes L, Lattanzio S, Moser BR, Stjepandić J, Wognum N (eds) Transdisciplinary engineering for resilience: responding to system disruptions. Advances in Transdisciplinary Engineering 16:302–311. https://doi.org/10.3233/ATDE210109
Zaoui F, Souissi N (2020) Roadmap for digital transformation: a literature review. Procedia Comput Sci 175:621–628. https://doi.org/10.1016/j.procs.2020.07.090

Chapter 6
Digital Safety in Rail Transport—Basic Assumptions

Adam Jabłoński

Abstract *Purpose*: The aim of this chapter is to present key assumptions to clarify the determinants of rail transport digital safety.

Design/Methodology/Approach: A critical literature review with theoretical and practical reflections aiming to explain the interdisciplinary nature of digital safety management in rail transport has been adopted as a methodology for this chapter.

Findings: This chapter identifies key theoretical approaches to understanding digital safety issues in relation to the life cycle of railway systems, as well as the role of software and hardware in digital safety management processes. Cybersecurity was also identified as a key challenge for today's railways.

Research limitations/implications: The limitations of the analytical and evaluation study result from the difficulty in identifying those issues, which are of the utmost importance when it comes to fully clarifying the issue of digital safety in view of a wide range of approaches to this issue.

Practical implications: The practical nature of the content described is manifested in critical analysis and examples and explaining factors responsible for rail transport digital safety.

Originality/Value: The originality of this chapter is an attempt to explain the relationship between safety and security aspects.

6.1 Introduction

The concept of safety has many variations, dimensions and interpretations. Based on this interpretation, digital safety is particularly important as an attribute of a modern and safe transport system. The digitalization of rail has become reality. Undoubtedly, the result of its application is not only a logical change in the process of creating solutions, but above all, ensuring railway safety at high speeds and higher quality parameters. The recipient of these principles is the entire society, whose expectations are extremely high and clearly defined, i.e. safe and convenient railways. To this end, the adopted and applied digital rail transport safety management mechanisms must be subordinated. These mechanisms are different and are based on diverse

A. Jabłoński and M. Jabłoński, *Digital Safety in Railway Transport—Aspects of Management and Technology*, Springer Series in Reliability Engineering,
https://doi.org/10.1007/978-3-030-96133-6_6

criteria from those used in analog railways with different dimensions of technological conditions. The skillful transformation of digital solutions in theory and practice without disruption for a safe, smooth and sustainable rail transport system is essential. Such thinking includes strategic, tactical and operational dimensions of the safety management of railway enterprises; they are expressed in this context in digital solutions adopted to ensure the safety of the rail transport system.

6.2 Theoretical Assumptions for Defining Rail Transport Digital Safety in Holistic Terms

The rail transport system is a cyber-physical system. Cyber-physical systems have emerged as a branch of research into embedded systems, focused in particular on the interaction between computational elements and physical entities (Wolf 2009). It is important to be aware that many safety–critical systems, such as those used to control rail transport, are inaccessible from the Internet, but have technically used equipment located in geographically distributed locations for control, power and telecommunications (Marrone et al. 2015). Then they can be widely used: Autonomous Cyber Physical Systems (ACPS) include industrial and advanced robots and autonomous navigation systems. Networked CPS or cyber-physical systems of systems (CPSoS) are large, distributed systems, e.g. smart grids and railway systems. Industrial automation and control systems (IACS) are also used to control physical processes in the oil and gas, nuclear industry, etc. (Bolbot et al. 2019).

Digital safety in this respect includes the components of the railway system that are directly based on digital platforms as well as those elements that are supported by digital solutions. Digital safety can be understood as the state of the railway system free of the risks arising from digital solutions implemented in that railway system, which directly or indirectly affect the probability of the occurrence of events causing human loss and/or damage. A condition for the ontological separation of digital safety from the general concept of railway system safety is that the concept of overall railway system safety is limited to those components that affect the emergence of risks in the context of digital solutions. Therefore, the concept of digital safety clearly narrows the scope for reflection on the safety of railway systems, drawing attention to the above-mentioned characteristics of the electronic and programmable areas of rail transport operation. Digital safety therefore applies only to situations where digital solutions exist to a significant extent and which, as a result of their very nature, can pose risks to railway safety. Wherever digital solutions do not exist at all or are marginally significant, digital safety does not exist. Therefore, in this discussion, digital safety therefore refers to relatively modern solutions in the railway sector where the share of digital solutions is high. Digitalization is crucial to increased accessibility, automation, connectivity and sustainable development of railway infrastructure. In this area, the solutions used should be safe. Attention should be paid to the distinction between the concepts of safety and security. Safety refers to the technical safety related to train traffic and is operational. Security, on the other hand, covers the protection of the entire railway system not only in terms of railway traffic, but also issues such as terrorist threats, hacking attacks, errors in the information flow security system and data-based decision-making processes. In terms of digital safety for fully digitalized systems, the concepts of safety and security are, in a sense, compatible and should be considered together. Process security in the flow of information between operators may be exposed to data protection risks and potential interference with processes and information flows. From this perspective, the concept of cybersecurity should be discussed in conjunction with the concept of safety, as a coherent concept that fulfills the assumptions of digital safety. It is also vital to pay attention to physical safety, especially if we think of the complex system of connections of devices built up within the full configuration of the railway system based on socio-technical aspects—in other words, the relationship between people and properly programmed railway traffic control devices. Physical safety is all that concerns the physical distribution of resources. The logic and efficient location of these resources in the design of railway systems is crucial. Therefore, this aspect should also be assessed in terms of safety criteria. Figure 6.1 presents the location of digital safety set against the background of a comprehensive approach to railway system safety. Digital safety covers both cybersecurity and process safety, which includes technical safety.

The diagram presents the concepts of functional safety, which is crucial to digital, physical and process safety. Safety, as stated in EN 61508-1:2010, is achieved by means of a number of systems that use diverse techniques, e.g. mechanical, hydraulic, pneumatic, electrical, electronic, and programmable electronic techniques. Therefore, each safety strategy must consider not only all the components of the separate system (e.g. sensors, control devices), but also all safety-related systems forming an overall combination of safety-related systems. In this way, all activities during the safety life cycle of systems which contain electrical and/or electronic and/or

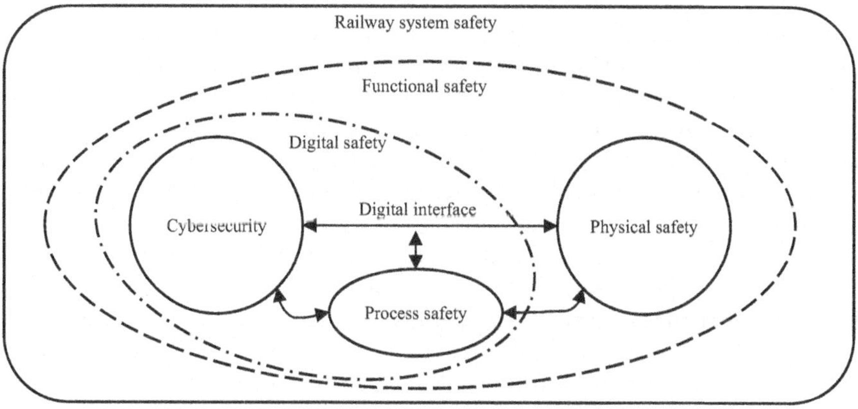

Fig. 6.1 Location of digital safety in the overall concept of railway system safety (*Source* Own study)

programmable electronic components (E/E/PE) are used to fulfill safety functions. All relevant phases of safety life cycles should be considered: total, an E/E/PE system and software (for example, from an initial concept through design, implementation, use, and operation to decommissioning) when E/E/PE systems are used to fulfill safety functions. Each phase of the overall safety life cycle must be divided into elementary activities, with the range, inputs and outputs specified for each phase. This standard is used to analyze E/E/PE safety-related systems and other risk mitigation measures so that the specification of E/E/PE system safety requirements related to safety can be defined in a systematic and risk-based manner. The holistic approach enables all phases of safety life cycles, namely total, an E/E/PE system and software, to be successfully completed. Within the scope of functional safety assurance, suitable procedures should be developed to ensure the handling of E/E/PE safety-related systems, including recommendations resulting from hazard and risk analysis, functional safety assessment, verification activities, validation activities, configuration management, incident reporting and analysis. Digital systems in terms of EN 61508-1:2010 consist of two fundamental areas of functional safety, namely safety related to the E/E/PE system life cycle and the software life cycle. The operationalization of the safety management process in the area of functional safety should be based on the concept of risk management. It is also necessary to identify hazards, hazard events and situations concerning end-user computing (EUC) and EUC control systems (in all types of work) in all reasonably foreseeable circumstances, including conditions of failure and reasonably foreseeable misuse. It is also necessary to establish a sequence of events leading to hazard events and to identify the EUC risk associated with hazard events. In particular, the development of a E/E/PE safety-related system use and operation plan to ensure that the required functional safety is maintained during its use and operation is essential. To illustrate the essence of functional safety, an example of the identification of total safety functions can be used, for example preventing the temperature increase in tank X above 250 °C and preventing the speed from exceeding 3000 rpm in the drive Y. This understanding of functional safety means that the subsequent obvious step is to identify a safety hazard. A vulnerability analysis must then be conducted to specify protection requirements. For each complete safety function it is necessary to specify the requirement of target safety integrity that will result in a tolerable level of risk. Each requirement may be quantifiably and/or qualitatively defined. In this way, a specification of the complete safety integrity requirements (EN 61508-1:2010) must be established. The approach to functional safety management is fully part of the area of digital safety management. The configuration approach considered in the life cycle of railway system components supported by the risk management concept, taking all phases of the use of those systems into account, seems complete when it comes to ensuring the expected level of safety. However, attention should be paid to technological progress, which poses a number of new challenges in terms of digital safety, as discussed further in the monograph. The treatment of digital safety in the context of functional safety determinants, given the popularity of the EN 61508-1:2010 standard, seems obvious.

6.3 Cybersecurity in Rail Transport and Technical System Safety—Safety and Security Aspects

Due to its dynamic digitalization, cybersecurity in rail transport plays an increasingly strong role. The adjustment mechanism is crucial here. In response to cyber threats, users change their behavior (Dodel and Mesch 2019). Relatively new threats of cyber-attacks are causing increasing concern about the safe operation of critical infrastructure, which includes railway infrastructure. When implementing the strategic provisions of key documents of the European Union, but also of global reports and legal acts, the task of cybersecurity protection is to establish procedures and obligations to report cybersecurity incidents to companies in key sectors, to establish specific security requirements for companies in key sectors, to adopt national strategies for network and IT system security, to set up a network of Computer Security Incident Response Teams (CSIRT), and to establish a special group which ensures strategic cooperation and information exchange, in particular considering the fact that cyber security incidents often affect many countries simultaneously. The establishment of the concept of the operators of essential services has allowed for the selection of companies from sectors such as banking, energy, transport, financial market infrastructure, digital infrastructure, health care, and drinking water supply and distribution. Such an approach at the strategic level implies action in the rail transport sector itself. As regards railways, the regulation covers key railway infrastructure managers, passenger railway undertakings and major freight railway undertakings. The scale of activity and market share is important here. The provisions also point to the role of the service infrastructure facilities as entities, which should also have a particular concern for ensuring security against cybersecurity threats. The way to cybersecurity protection in the context of the conditions of functioning of the entities is the establishment, by the operator of essential services, of a security management system in the information systems used to provide the essential service, ensuring in particular:

1. The systematic estimation of the risk of an incident and managing the risk of an incident based on the estimated risk;
2. The implementation of appropriate and risk-proportionate technical and organizational measures, taking into account the state of the art and ensuring the security of the information systems used to provide an essential service, including:

   - maintenance and safe operation of information systems,
   - physical and environmental safety, including access control,
   - security and continuity of supply of services, which the provision of the essential service depends on,
   - the implementation, documentation and maintenance of action plans which enable the continuous and uninterrupted provision of the essential service, as well as its confidentiality, integrity, accessibility and authenticity,

- the monitoring of information systems used to provide the essential service on a continuous basis.

3. The collection of information on cybersecurity threats and the incident vulnerabilities of information systems used to provide the essential service;
4. Incident management;
5. The application of measures to prevent and mitigate the impact of incidents on the security of the information systems used to provide the essential service, including:

   - the use of mechanisms which ensure the confidentiality, integrity, accessibility and authenticity of data processed in information systems,
   - attention to software updates,
   - protection against unauthorized modification in information systems,
   - taking immediate action following the perception of cyber vulnerabilities or threats;

6. The use of communication means for proper and secure communication within the framework of a national cybersecurity system.

The operator of the essential service is also obliged to ensure that periodic security audits of the information systems used to provide the essential service are conducted. It is also important to have an IT system that helps to provide notification of the deadline for taking verification or corrective actions in relation to engineering structures on the railway network. Building information modelling (BIM), i.e. modelling information about building assets that include railway buildings, may be helpful.

Cybersecurity is therefore already a highly formalized concept based on regulations and standards developed. The high awareness of threats, responsibility of managers and the level of training of railway system operators to skillfully respond to crisis situations are a condition of a proper approach to these problems. Protection against cyber-attacks on technical systems is currently one of the crucial issues for the construction of the safety management systems of railway undertakings and infrastructure managers which, as mentioned above, meet the definition of the operators of an essential service. The identification of cyber security threats, despite being part of the concept of security rather than safety, seems to need to be considered jointly in the age of the digitalization of railway systems. In fully digital technical environments, these concepts appear to be complementary, and in the future such ontological division is likely to be eliminated and differences in this respect will be blurred. After all, any risk pertaining to the functioning of the railway system affects the level of safety, and the artificial separation of these two concepts as separate ontological entities at the current stage of the development of the digital transformation of the railway systems does not seem to be justified.

6.4 The Life Cycle of Railway Systems and the Digital Safety of Technical Systems

The life cycle of railway systems is a crucial determinant of the assessment, verification and validation of safety solutions. Any electronic and programmable solution must be considered in terms of the life cycle of the technical system. Each technical system is characterized by a certain service life and can generate different hazards during operation, which may result from both intrinsically relevant characteristics and parameters, as well as performance and functions. The life cycle of a digital system should be understood as a defined concept of the distribution over time of the activities taking place during the development, production and operation of a specific type of system. This cycle covers the period from the emergence of the user's need to build a digital system, to the presentation of its ideas, construction, and use, adapted to possible changes in operation, and ending with its decommissioning. At each of these stages, the expected level of safety should be ensured through safety management activities. Within the life cycle of railway systems, which are designed to be safe, the concept of the overall safety life cycle must be isolated as a technical framework for the systematic treatment of activities required to ensure the functional safety of E/E/PE safety-related systems. To this end, it is reasonable to apply the provisions of the EN 61508-1:2010 standard. The overall safety life cycle includes the following measures to achieve tolerable risks, namely E/E/PE safety-related systems and other risk mitigation measures. The overall safety life cycle includes several crucial stages, namely the life cycle of the digital system at the implementation stage, the life cycle of the software at the implementation stage, the relationship between the overall safety life cycle and the safety life cycles of E/E/PE and software. The stages of the safety life cycle should include issues that are sequentially implemented in the project, such as the complete definition of the scope of analysis, hazards and risks, total safety requirements, the allocation of total safety requirements, complete operational and maintenance planning, complete safety validation planning, complete installation and commissioning planning, specification of E/E/PE safety requirements, the implementation of E/E/PE safety-related systems, other risk mitigation measures such as specification and implementation, complete installation and commissioning, complete safety validation, total operation, maintenance and repair, complete modification and renewal (EN 61508-1:2010). The digital safety of technical systems will be safe if all the above stages of the safety assessment for the different stages of the technical system life cycle are conducted and documented correctly. The correct implementation of the life cycle stages of railway systems in terms of safety means the reliability of inspections conducted, their verifiable verification, the documentation of full identification and identifiability of events, management supervision, compliance with relevant procedures, as well as responding to incompatibility and proactively seeking compatibility of railway systems through the use of methods of the monitoring and audits of safety systems. It will also be important to assess technical compatibility and safe integration in the case of changes in railway systems during their life cycle. It should also be borne in mind that problems may occur during the various stages and that it will then be necessary to return to previous stages, but the probability of such situations should be minimized. In the case of the digital system design process, the logical consistency of the sequential and iterative stages must be ensured with an indication of milestones. To this end, the requirements supported by the definition of objectives and the safety requirements of the technical systems should be clearly defined. In the next step, the system needs to be thoroughly analyzed and a logical model of the system needs to be built. As part of the design, conceptual assumptions should be built to determine the detailed parameters of the project. Further implementation solutions related to testing individual project modules in a specific environment of the technical and operation ecosystem based on programming solutions should be implemented. A series of tests should be conducted further to assess the compatibility of the configuration of the components to be linked. The final phases of the system design will include implementation, taking user participation into account. Once the user's comments have been approved, system maintenance conditions should be established, and modifications, alterations and extensions of the designed system functions should be implemented if necessary. For these types of projects, the iteration-based approach is the leading one. For this purpose, Modified Cycle V in the rail transport sector, is used, the modified version of which in terms of digital economy solutions is presented in Fig. 6.2.

This model specifies the full design cycle, including the operational stages and the decommissioning of the railway system. Such a highly formalized method is justified to ensure the repeatability of the process of handling the systems introduced on the railways and to ensure their proper and safe operation. In the case of systems based on digital solutions, these stages are also particularly important in terms of the verification of interfaces between digital devices, but also in specific cases, with

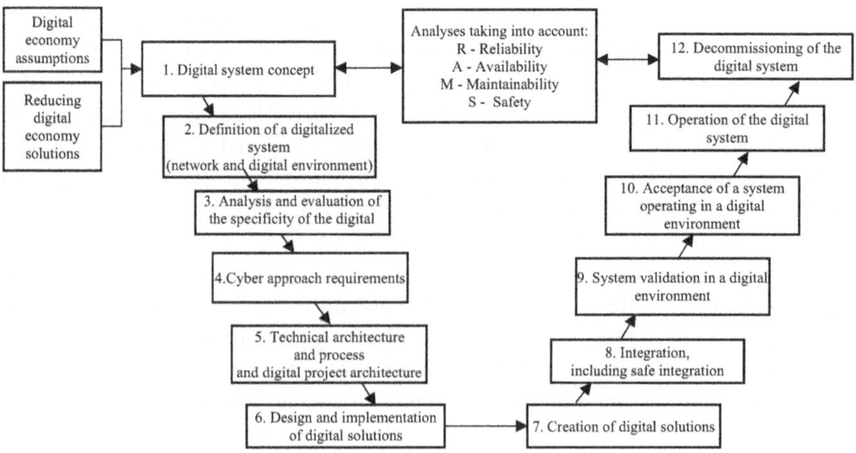

Fig. 6.2 Modified Cycle V according to EN 50126-1 (*Source* Own study based on EN 50126-1, Railway Applications—The Specification and Demonstration of Reliability, Availability, Maintainability and Safety (RAMS)—Part 1: Generic RAMS Process)

systems with a different technical concept from outside the digital economy solution package.

6.5 Identification, Design and Construction of Safety Criteria in Railway Hardware and Software Solutions

The process of designing railway systems and equipment in the context of the safety criterion requires the identification of specific safety criteria. Electronic safety-related signaling systems cover hardware and software aspects. Safety-Related Application Conditions (SRACs) are an essential prerequisite for the development of safety-related systems. According to the provisions of the EN 50129:2019 standard, when using existing equipment which only performs part of the safety function, the equipment should be properly integrated into technical and procedural safety management. To this end, certain requirements apply; in particular, a minimum set of information is available such as functionality, interfaces, hardware and/or software limitations, equipment failure rate, environmental conditions and other conditions of use. A crucial aspect of ensuring the expected level of railway safety is the proper management of the configuration constituting a structured approach and the combination of the system hardware and software for its intended use, taking into account the fulfilment of safety criteria. In terms of safety, the relationship between hardware and software should be strictly managed, as detailed in EN 50128. Therefore, software must be integrated with the target equipment and the functionality must be verified to make sure the solution being designed is safe. The purpose of this process is to determine whether the software, which may include the process, documentation, system, subsystem hardware and/or software components, fulfills certain requirements, which should be described in the relevant structured evaluation: is the software relevant for the purpose and does it achieve the expected safety characteristics of the railway system? In this situation, the appropriate level of error tolerance built in as the ability of the system to ensure the continuous proper provision of services according to the specification, in the presence of a limited number of hardware or software failures, corresponds to the requirements of software and hardware configurations in the context of the design of safe railway systems. Thus, it is necessary to apply a system safety integrity assessment process, which indicates the required degree of certainty that an integrated system consisting of hardware and software will satisfy certain safety requirements, which is a priority and guideline for technical system designers based on safety criteria.

6.6 Conclusion

Digital safety poses a new challenge for digitalized technical railway systems. Digital safety should be based on requirements specified in the relevant standardization standards—CENELEC standards. As not all digital solutions can be classified as those for which such standards should be applied, there is a need for a broader analysis of this issue. A significant number of digital solutions do not have a direct impact on railway traffic safety, but their share in the overall railway system and the supply chain also creates risks on their side. Cybersecurity is one of those challenges where theory and practice do not have enough experience to recognize that all related hazards are known and defense mechanisms against hacker attacks are fully developed. New forms of attacks—and defense against them—are continually being designed. All such risks may be subject to a functional safety procedure, which is understood as a general approach to all activities in the safety life cycle of systems containing electrical and/or electronic and/or programmable electronic components. Extending this approach to other systems supporting the functioning of the railway system seems right and justified. Digital safety is a major challenge, as many digital and hybrid interfaces pose a significant risk to the expected safe continuity of operation of railway systems.

References

Bolbot V, Theotokatos G, Bujorianu LM, Boulougouris E, Vassalos D (2019) Vulnerabilities and safety assurance methods in Cyber-Physical Systems: a comprehensive review. Reliab Eng Syst Saf 182:179–193. https://doi.org/10.1016/j.ress.2018.09.004

Dodel M, Mesch G (2019) An integrated model for assessing cyber-safety behaviors: how cognitive, socioeconomic and digital determinants affect diverse safety practices. Comput Secur 86:75–91. https://doi.org/10.1016/j.cose.2019.05.023

EN 50126-1, Railway applications—the Specification and demonstration of reliability, availability, maintainability and safety (RAMS). Part 1: Generic RAMS process.

EN 50128/A1:2020, Railway applications—communication, signalling and processing systems. Software for railway control and protection systems, 2011.

EN 50129, Railway applications—communication, signalling and processing systems. Safety-related electronic systems for signalling, 2019.

EN 61508-1:2010, Functional safety of electrical/electronic/programmable electronic safety-related systems. Part 1: General requirements.

Marrone S, Rodríguez RJ, Nardone R, Flammini F, Vittorini V (2015) On synergies of cyber and physical security modeling in vulnerability assessment of railway systems. Comput Electr Eng 47:275–285. https://doi.org/10.1016/j.compeleceng.2015.07.011

Wolf W (2009) Cyber-physical systems. Computer 42(3):88–89. https://doi.org/10.1109/MC.2009.81

Chapter 7
The Creation of Digital Innovation in Rail Transport

Adam Jabłoński

Abstract *Purpose*: The purpose of this chapter is to present key assumptions for the design of digital innovations in rail transport in terms of rail transport safety.

Design/Methodology/Approach: A critical literature review with theoretical and practical reflections aiming to present the interdisciplinary character of the creation of digital innovations in rail transport has been adopted as a methodology for this chapter.

Findings: This chapter identifies crucial theoretical and practical approaches to understanding the issue of creating digital innovation in view of the difficulty of placing these solutions in service in the rail transport sector.

Research limitations/Implications: The limitations of the analytical and evaluation study result from the difficulty in identifying issues that are crucial to fully explaining the testing, verification and validation of innovative solutions in the digital economy.

Practical implications: The practical nature of the content described has been focused on critical analysis and examples and explaining factors responsible for shaping innovative digital railway solutions.

Originality/Value: The originality of this chapter may be an attempt to identify crucial problems related to placing digital, innovative solutions dedicated to rail transport in service.

7.1 Introduction

When considering railway enterprises in terms of innovation, they can certainly be called innovative traffic organizations (Ljubaja and Mlinarić 2019). This cognitive perspective opens up a wide range of possibilities that can be used in dynamic digital and technological rail transport systems. Innovation is primarily intended to change, improve and increase parameters, as well as improving functions and performance. Such an understanding of innovation makes it not only a creator of change, but above all a catalyst for the new dimension of railway efficiency and effectiveness, including in terms of the especially important criterion of its safety. Innovations and the types thereof range from open to closed innovations, from technological to

A. Jabłoński and M. Jabłoński, *Digital Safety in Railway Transport—Aspects of Management and Technology*, Springer Series in Reliability Engineering,
https://doi.org/10.1007/978-3-030-96133-6_7

organizational ones, and from process to product ones. Each has a unique place and role in rail transport systems. They affect numerous organizational, technological and operational parameters of railways, and an important attribute thereof is their high productivity, measured by their effectiveness and efficiency. In this dimension, innovation is seen as a stimulating and driving factor of railway development, mainly in the dynamic digital transformation of railways.

7.2 Innovation and the Specificity of Rail Transport

Theoreticians and practitioners make extensive use of digital transformation to understand how existing companies may need radical transformation to succeed in the emerging digital world (Nambisan et al. 2019). The digital revolution is underway, which changes the way we move and produce. Success in the digital revolution means that the railway industry must benefit from the best available human-centered technologies (Carranza et al. 2020).

In the digital environment and current conditions, the competitiveness of the railway industry is determined by the level of implementation of innovative solutions and the share of the use of intelligent control systems (Sirina and Yushkova 2021). In order to improve rail transport safety as well as the effective and efficient improvement of logistics processes in rail transport, the use of organizational, technical and operational solutions using various types and categories of innovation is required. Innovation plays a particular role in the rail transport sector. Nowadays, the development of this sector is determined by digital innovation. Investment in high-speed rail is an important determinant of innovation in rail transport. This generates, inter alia, a significant impact of infrastructure on innovative cooperation between enterprises at different locations aimed at the deployment of high-speed rail (HSR) (Hanley et al. 2021). Previous studies show that the role of HSR in the city, economy and society has two sides. On the one hand, the launch of HSR improves the regional advantages and competitiveness of cities or regions; on the other hand, it promotes economic growth that is conducive to the perception of the expansion of the economy into large cities, thereby promoting the sustainable development of regions (Yang et al. 2021). Empirical results imply that the launch of HSR significantly improves the performance of urban innovation; there is, however, considerable spatial heterogeneity. The launch of HSR not only promotes agglomerations and urban expansion factors, but also strengthens interurban economic links, thus improving the performance of urban innovation through the externalities of the agglomeration and networks. However, compared to the internalities of agglomerations, the external factors of the network have a greater impact on the city and the efficiency of innovation (Tang et al. 2021). According to this interpretation, on the one hand, innovation in rail transport has a significant impact on rail transport safety, but on the other hand, it is important for achieving higher rates of operational readiness of technical and operating systems. Improvements also concern the better use of the infrastructure resources of the transport system, which results in the better parameterization of the results achieved, including higher capacity, higher speeds, elimination of bottlenecks in the transport system, etc. Innovation also significantly contributes to improving railway communication systems, as evidenced, inter alia, by the shift from analog to digital communication systems.

7.3 Problems and Barriers to Placing Innovative Solutions in Rail Transport into Service

The application of various solutions which are relevant to the rail transport sector requires the adoption of established and documented, regulated and standardized rules of conduct. This significantly prolongs the process of placing innovative solutions in rail transport into service. It must be made clear that the procedures for placing innovative rail transport solutions into service are not short. This results primarily from obtaining certainty that the solution will not only fulfill its functions but, above all, that it will be safe for the entire railway system. This means that this solution must meet the requirements of technical compatibility and safe integration. In order for this to be possible, innovative solutions need to proceed in a way which is typical of effective and efficient project management mechanisms, with review, verification and validation expectations being fulfilled. This requires so-called piloting, which ensures that test results that are sufficient to decide on the safe operation of these solutions are obtained. Digitalization-based innovative solutions are increasingly being implemented for use in rail transport systems. The basic documents authorizing these solutions are the Technical Specifications for Interoperability. The fundamental commonly implemented system is the ETCS. The use of the ETCS as overriding to the national railway traffic control devices of the basic layer is intended to ensure, on the one hand, the interoperability of control solutions in the European Union, and on the other hand, a high level of railway traffic safety without interfering with the base layer systems used in individual EU countries. This provides an opportunity to gradually digitalize railways in the context of the development of ETCS itself as well as computer applications supporting large data collection and management systems. Innovative solutions of this type will have to undergo complex certification schemes implemented by International Industry Organizations such as the European Union Agency For Railways and/or the National Safety Authority in individual countries. Similarly, innovative solutions are created around the Announcement Communication System using GSM networks. All these solutions need to undergo a number of studies, analyses and tests. For example, compatibility assessments should be conducted for the deployment of ETCS and GSM-R solutions, which are additional inspections of ETCS and GSM-R systems, as these are necessary to achieve the highest possible level of confidence in the technical compatibility of on-board equipment with the individual trackside deployments of the system. These inspections underlie further harmonization of subsystems. For ETCS, these are ETCS System Compatibility tests which aim to demonstrate the technical compatibility of on-board equipment with the ETCS trackside part. For GSM-R, these are Radio System Compatibility (RSC) tests which aim to demonstrate the technical compatibility of a train radio or EDOR with GSM-R track-side equipment. ESC and RSC compatibility tests are intended to facilitate the operation of new subsystems, as their proper and efficient functioning in a given environment is fully verified at the stage of ESC and RSC compliance demonstration. The results of the ESC/RSC tests are checked by the notified body each time. Based on the results of the ESC/RSC tests and the inspection carried out by the notified body, the applicant issues a declaration of ESC/RSC compatibility which forms part of the EC declaration of subsystem verification. The complexity of placing railway vehicles and CCS control command and signaling systems into service is related to the assessment of technical compatibility and safe integration. In each case, compliance with both parameters needs to be demonstrated. Technical compatibility means the ability of two or more structural subsystems or parts thereof which have one or more common interfaces to interact with each other while maintaining their individual designed operating state and expected performance level. Establishing technical compatibility with the network means verifying and documenting, in the technical documentation accompanying the EC declaration of verification, the parameters of the vehicle type that are relevant to technical compatibility with the network and, where applicable, compliance with the limit values defined for that network. The parameters include physical and functional features, and the verification must be conducted in accordance with the rules of the network (2014/897/EU). Safe integration means action to ensure that the integration of a component (e.g. a new vehicle type, network design, subsystem, part, component, software, procedure, and organization) into a larger system does not cause unacceptable risks for the system overall (European Union Agency for Railways 2020). Safe integration includes relationships between different systems and technologies, including both older and newer ones. A relationship also occurs between parts suppliers—elements and components, different ways of maintaining these parts (subsystems, components), different rules of railway system operation, different levels of system complexity, different levels of system safety and other issues. Safe integration will focus on assessing how the interface connection affects the safety of the entire connected system and the impact of this connection on all aspects of safe integration at all stages of the life cycle of the designed and changed technical system. These assumptions impose extremely high demands on innovative solutions. Innovation must therefore, over time, be standardized and adapted to the rules of the rail transport sector. Attention should also be paid to those innovative digital applications which are not subject to such strict restrictions on placing them in service. These will be supportive applications, the functionality of which can be infinite. It is important that they fit into the system of base equipment and technical systems which, as with those related to railway safety, fulfill all the requirements for certification and placing them in service.

7.4 Technical, Operational and Organizational Innovations and Rail Transport Safety

The problem of capturing value for innovators in the digital economy raises certain other challenges than those in the industrial economy. This inevitably requires an understanding of the dynamics of the platform and specific ecosystems (Teece 2018). In this case, the railway ecosystem, its features and conditions pave the way for the creation and implementation of different types of innovation. Technical innovations are related to the use of a given instrument that is relevant for the engineering/technical attributes of the systems concerned of an object-oriented nature. Operational innovations are process-driven due to the conditions of railway traffic management and the associated other dynamic processes. Organizational innovations are the result of management decision-making processes. It is important that these innovations complement, rather than exclude, each other. Therefore, the determinants of a socio-technological system such as a railway system should be perceived in a holistic way. Only then will a multitude of proposed and implemented solutions be appropriately selected, and only such solutions that will produce symbiotic and synergistic effects will be selected. Technical, operational and organizational innovations are characterized by different specificities and interpretations. In the digital economy, this will have a different approach to defining differentiating features. Technological innovations in the field of digitalization are such innovations that are based on new value delivery formulas resulting from specific technical specifications and new opportunities for railway system users. At the same time, they have the potential to replace existing solutions and provide innovative formulas for the implementation of technical railway operations. Operational innovations mean the type of innovations that concern the operation conditions of railway systems at different stages of their life cycle. The optimization of operational parameters is their main goal. An innovative system according to the operation criterion should be characterized by a high level of reliability, maintainability, availability and safety. In a word, it should fulfill the requirements of the RAMS model, which is an analysis of installations allowing for the determination of possible design or modernization options in terms of their reliability, availability, maintainability and safety.

RAMS allows for the identification of potential bottlenecks, the performance of a preliminary analysis of the criticality of equipment, the estimation of the availability of installations, the pre-determination of planned and unplanned traffic maintenance activities, the development of an optimal stoppage strategy, the analysis of the impact of the size of maintenance personnel, the development of strategies to mitigate the effects of anticipated damage, and the optimization of reliability at the design stage by analyzing a wide range of options. Operational innovations, like other types from this division, will be compared with certain standards and rules which are relevant to the rail transport sector. Organizational innovations, in turn, will relate to the conditions for the implementation of work organization processes, especially in the context of the work processes of employees who perform tasks critical to railway traffic safety. They will address methodologies adopted for monitoring work responsibility, as well as supervisory and communication processes. Innovation in this area will have a wide impact on the potential for establishing a positive safety culture in railway enterprises. All these innovations significantly influence the processes of rail transport digital transformation and the full integration of all digital solutions from the areas pertaining to safety and security.

7.5 Testing, Verification and Validation of Innovative Digital Solutions in the Context of Safety Criteria

In order to be able to apply different innovative solutions in rail transport, multiple validations thereof are necessary, i.e. those which provide direct proof that the solution fulfills the requirements for its intended use. In order to do this, a certain logic must be developed which is suitable for achieving the intended results. On the one hand, there is a need for appropriate technical compatibility to enable a standardized use of a particular solution in a given system or technical subsystem, and on the other hand, safe integration which ensures the safe operation of the system, subsystem, or component in a complex, comprehensive technical system. The way to achieve this type of assumption can be the testing, verification and validation of mainly innovative digital solutions. Testing is part of verification and validation. Verification in this case concerns the provision of feedback pertaining to whether the input data at the different stages of testing are still suitable in order to achieve the assumed outputs evaluated in the process of further multiple validation. It should be noted that only the correct sequence between testing, verification and validation determines the subsequent safe operation of an innovative solution. Unfortunately, improper and erroneous conduct in this area can result in a railway disaster, a major accident or a railway accident. Therefore, the mapping of reality in the processes undertaken in the testing ground should be as precise and structured as possible in the procedures adopted. Specific assumptions and requirements relate to innovative digital solutions. The concepts of review, verification and validation are derived from the assumptions of quality management systems. Any process of designing development solutions should be subject to such assessments. The review answers the question of whether the input data is consistent with the output data in the process of implementing the change project, i.e. a technically compatible and safe subsystem satisfies the output data after the implementation of the change project (the assumption has been achieved). So here all the evidence will indicate that what was assumed was built or done. Project verification aims to provide evidence that the product fulfills the defined requirements, i.e. whether the change project fulfills all essential requirements pertaining to technical compatibility and safe integration in the railway system. The evidence consists of documents such as regulations and requirements in this area. Instead, the project validation will consist of verifying the correctness and providing evidence that the outcome of the change project (a modified subsystem) fulfills the user's needs and requirements. Therefore, all test results confirming that the change ensures safe operation of the subsystem must be included here. The presented way of assessing innovation and changes in railway systems concerns both digital applications and analog solutions. In any case, these milestones must be conducted. Such an approach ensures that all components, parameters including functions and performance are obtained, and the designed systems and the changes thereto are confirmed as being capable of safe operation.

7.6 Innovation and Exchange of Information in Railway Systems—Subjective and Objective Perspectives

The exchange of information is crucial in rail transport processes. These processes occur in several key layers and areas of activity of railway market operators within their core operational tasks, namely:

- A layer of structural subsystems—infrastructure, energy and control—trackside and on-board equipment, rolling stock.
- A layer of functional subsystems—railway traffic, telematics.
- A layer of maintenance and predictive maintenance of railway infrastructure and railway vehicles.
- A layer of rail transport logistics taking into account, inter alia, the following tasks: the technological and IT solutions of systems for railway traffic management on railway lines, at railway stations, transshipment terminals, railway sidings, other railway transport facilities providing full transport, maneuvering and handling services.
- A layer of communication based on the architecture of data exchange and sharing in rail traffic, for the purposes of synchronizing the activities of terminals, customs and revenue services, control companies, undertakings and railway operators, infrastructure managers in the field of Local Control Centers, Train Dispatcher Assistance Systems and Operation Register Systems, as well as other operating systems necessary for the handling of passenger and freight traffic.

In addition, the following issues are often digitalized:

- Processes of train/train composition/wagon notification by railway operators in transshipment terminals,
- Processes of freight forwarding notification of loading/unloading wagons in terminals,
- Handling processes in transshipment terminals,
- The process of verification of empty and full wagon formations, including technical condition, scanning, number reading, damage identification and weighing,
- The processes of rail transport service by railway and intermodal operators,
- The process of rail transport service by en-route and maneuvering railway undertakings,

- Railway traffic control processes by infrastructure managers in en-route transport services to/from land terminals,
- Processes for managing the use of railway infrastructure by infrastructure managers at railway stations according to plans for train/train composition/wagon notification in transshipment terminals,
- Processes for handling rail freight by customs and revenue services, border inspections and control companies,
- Processes for monitoring and managing railway traffic on railway lines and transshipment terminals,
- Processes for providing information about incidents on tracks, repairs and investments, and exclusions of railway infrastructure sections,
- Integrated planning, confirmation and monitoring of train arrivals/departures at railway stations,
- Integrated planning, confirmation and monitoring of train drives, making wagons available and receiving them at terminals, wagon stops, track and siding occupancy,
- Integrated information on the carriage of hazardous goods,
- Communication, exchange and sharing of data and documents in the operational activity of all rail transport users,
- In the case of intermodal transport, the area of digital innovation may include data monitoring and decision making on issues such as: details of containers for unloading, container number, position of the container on the wagon together with the wagon number, ISO type, seal numbers, container weight, hazard class, organizer—ordering party who bears the handling fee for rail service, place of dispatch, and so on.

There is space for digital economy solutions everywhere. The level of application of digital solutions to these areas of operation by rail transport operators is extremely broad and can soon lead to the full digital transformation of this sector of the economy. Innovation in this sector is an inherent feature of its progress.

7.7 Conclusion

Innovation is the cornerstone of digital solutions in the rail transport sector. The digital economy stimulates innovation based on new forms of action. The rail transport sector, as a sector highly regulated by law, is characterized by a number of restrictions pertaining to placing products in service. In particular, innovative solutions are subject to a number of studies, tests and evaluations. To a large extent, these studies and analyses must address safety issues. Innovators must not only strive to create new functionalities based on innovative solutions, but first of all prove that the proposed solution is safe. The creation of innovation in terms of the safety criterion is a fundamental problem for these conditions to be solved.

References

Carranza G, Garcia M, Sanchez B (2020) Activating inclusive growth in railway SMEs by workplace innovation. Transp Res Interdiscip Perspect 7:100193. https://doi.org/10.1016/j.trip.2020.100193
Commission Recommendation of 5 December 2014 on matters related to the placing in service and use of structural subsystems and vehicles under Directives 2008/57/EC and 2004/49/EC of the European Parliament and of the Council (2014/897/EU)
European Union Agency For Railways (2020) *Clarification note on safe integration 1.0,* 1209/063
Hanley D, Li J, Wu M (2021) High-speed railways and collaborative innovation. Reg Sci Urban Econ 103717. https://doi.org/10.1016/j.regsciurbeco.2021.103717
Ljubaja I, Mlinarić TJ (2019) The possibility of utilising maximum capacity of the double-track railway by uing innovative traffic organisation. Transp Res Procedia 40:346–353. https://doi.org/10.1016/j.trpro.2019.07.051
Nambisan S, Wright M, Feldman M (2019) The digital transformation of innovation and entrepreneurship, progress, challenges and key themes. Res Policy 48(8):103773. https://doi.org/10.1016/j.respol.2019.03.018
Sirina N, Yushkova S (2021) Polygon principles for integrative digital rail infrastructure management. Transp Res Procedia 54:208–219. https://doi.org/10.1016/j.trpro.2021.02.066
Tang C, Guan M, Dou J (2021) Understanding the impact of high speed railway on urban innovation performance from the perspective of agglomeration externalities and network externalities. Technol Soc 67:101760. https://doi.org/10.1016/j.techsoc.2021.101760
Teece DJ (2018) Profiting from innovation in the digital economy: enabling technologies, standards, and licensing models in the wireless world. Res Policy 47:1367–1387. https://doi.org/10.1016/j.respol.2017.01.015
Yang X, Zhang H, Lin S, Zhang J, Zeng J (2021) Does high-speed railway promote regional innovation growth or innovation convergence? Technol Soc 64:101472. https://doi.org/10.1016/j.techsoc.2020.101472

References

Chapter 8
Emerging Modern Trends and Technologies Increasing the Level of Rail Transport Safety

Marek Jabłoński

Abstract *Purpose*: The aim of the chapter is to present the key assumptions for identifying emerging modern trends and technologies which increase rail transport safety.

Design/Methodology/Approach: The methodology for this chapter is a critical literature review with theoretical and practical reflections which aim to identify trends in the development of the digitalization of rail transport that have the greatest impact on railway safety.

Findings: This chapter identifies key technical solutions that significantly change the approach to understanding digital safety in rail transport.

Research limitations/implications: The limitations of the analytical and evaluation study result from the difficulty in identifying those issues that are of the utmost importance to fully explaining the directions for the development of digital economy solutions that improve rail transport safety.

Practical implications: The practical nature of the content described has been focused on critical analysis and examples and the identification of the most important solutions currently used in rail transport for safety management focused on the use of technological solutions supporting the achievement of a high level of safety.

Originality/Value: The originality lies in the broad perspective of these rail transport digitalization technologies in the context of safety. The multitude of solutions proves the phenomenon of rapid technological progress in this area.

8.1 Introduction

Technologies change the world and social behavior; they revolutionize the economy and the thinking of entire communities, both local and global. A complete technological and digital revolution is taking place at this point in time. Robots are replacing people, technical systems limit the human factor to a minimum, and analog technologies are turning into digital ones. Safety based on analog solutions is de facto ceasing to exist. Requirements as well as criteria, rules and methods are changing. The railway ecosystem has a new development space and a new dimension of the

A. Jabłoński and M. Jabłoński, *Digital Safety in Railway Transport—Aspects of Management and Technology*, Springer Series in Reliability Engineering,
https://doi.org/10.1007/978-3-030-96133-6_8

market, finance and safety. Digital ecosystems are powered by new digital technologies that have transformed and changed the very nature and scope of traditional interdependencies. Digital ecosystems are increasingly expansive with their reach and importance beyond traditional value chains and the logic of the functioning of conventional industries (Subramaniam et al. 2019). The multitude of technological trends and digital solutions based on the so-called new economies occurs in this dimension. Even their systematization is a great challenge, let alone their logical and justified use. In this interpretation, the appropriate selection—which is relevant to their application in rail transport—is crucial. This is not an easy task; this reasoning should be driven by the wide application of these solutions in numerous areas of rail transport, seen from the point of view of a safety criterion.

8.2 Class a Control Systems as a Determinant of Technological Development and Railway Safety

The most popular train control system in Europe is ERTMS, which provides a digital solution that ensures the interoperability of rail transport, i.e. the possibility of free movement of trains on national railway networks (infrastructure owners) without having to stop at the borders and replace locomotives or train drivers. Within the framework of this solution, three levels are proposed, which define an ever-higher level of guaranteed safety. These are ETCS Level 1, Level 2 and Level 3.

In terms of securing the train by means of speed control, the ETCS Level 1 system, which is intended to be used on secondary and poorly loaded lines, where international trains are not driven, is compatible with this functionality, as at ETCS 1 a train driver must observe signals positioned at the tracks. ETCS Level 1 is an initial level that guarantees the safe passage of the train. Its task is to ensure that the train does not travel beyond the boundary of the established and secured block sections on the route and that it does not exceed the permitted speed on any section of the route. From the range of technological solutions, ERTMS/ETCS Level 1 provides for three options of equipment. Level 1 without updating is based on the transmission of driving authorizations to the vehicle by means of light signals. The signaling device is accompanied via the interface by a switchable balise on the track, which transmits permission to drive depending on the indication of the signal. The signal transmitted by the balise is received by vehicle antennas and processed in the cab to visualize driving information to the train driver and supervise his work (Dąbrowa-Bajon 2007).

Level 1 with additional balise upgrades differs from the previous configuration in that additional balises are installed on the route to the semaphore as a means of updating the driving information. This gives the opportunity to increase traffic capacity and save energy consumption by updating information more accurately than "through the eyes of the train driver." The disadvantage of such a solution is the increase in line equipment, including the number of cables laid along the line.

Level 1 with a loop upgrade in the track allows for the continuous updating of information at the section approaching the semaphore. This configuration allows for further increase in the line capacity and reduced energy consumption by providing better updates than "through the eyes of the train driver" over a longer distance. It also allows one to opt out of trackside semaphores (Dąbrowa-Bajon 2007).

The use of any of the above-mentioned configurations on the line allows vehicles equipped (for line drives) to automatically adjust to higher levels of ERTMS. An advantage of such a system is the achievement of technical interoperability requirements, while a disadvantage is the impossibility of achieving operational interoperability (required for lines of international importance) (Gradowski 2003).

In Level 1, it is also possible to update the information by means of the GSM-R system, allowing the automatic matching of traction vehicles equipped with higher levels of the ERTMS. This configuration allows for the achievement of operational interoperability, as well as maintaining capacity. The use of this hardware makes it possible to forgo light signals only in the case of homogeneous traffic. Another possible hardware configuration is level 1 with an update by the Specific Transmission Module (STM). Updating information sent by the STM to the national *track-to-vehicle* transmission system is used by track-side equipment at the expense of another system. This configuration ensures that technical interoperability is achieved. However, when using this configuration, it is necessary to install an on-board STM on vehicles (on a line equipped with ETCS) which are equipped with a higher level of ERTMS equipment. It is also necessary to install a national on-board STM for the national system on the vehicles of other undertakings entering the neighboring railway network. Many of the features of this solution depend on a national system that would be used for the transmission of *track-to-vehicle* information (Gradowski 2003).

Level 2 is when the track circuits transmit a track occupancy and maximum speed signal to the locomotive, while semaphores themselves may not be installed only if all trains running on the line are equipped with ETCS. From a technical point of view, the ETCS 2 level is the traffic control based on continuous two-way data transmission. A traction unit (in addition to the level 1 locomotive equipment) must additionally be equipped with digital radio channel equipment (EURORADIO). Balises placed in the tracks are the basis for the location of vehicles, and the track is additionally equipped with RBC (Radio Block Center) radio control centers that communicate with the locomotive. Level 2 does not require the train driver to know the signaling applicable to the railway. It is intended to be used primarily on international and high-speed lines.

Level 3 is implemented in such a way that the task of controlling track-side occupancy is transferred from trackside equipment to vehicle equipment. This allows for the independence of block sections and allows track circuits and axle counters to be forsaken. The level 3 locomotive (in addition to the locomotive level 2 equipment) must additionally be equipped with a train integrity unit. Level 3 significantly increases the capacity of the railway line on which it is installed.

It is possible to move from level 1 to 2 and from 2 to 3. ETCS levels 2 and 3 are also called ATC (Automatic Train Control). ETCS is key to dynamizing the needs of the digitalization of the railway system and poses a crucial challenge by stimulating significant progress in the field of railway safety.

8.3 Artificial Intelligence and its Use in Improving Railway Safety

Since the 1950s, when McCarthy introduced the term "Artificial Intelligence" (McCarthy 1959), the field of artificial intelligence (AI) has developed in two dimensions: human focus and a rationalist approach (Borges et al. 2021). It is important to be aware that two crucial levels of AI are currently being developed:

1. Artificial Narrow Intelligence (ANI),
2. Artificial Super Intelligence (ASI).

Artificial Intelligence (AI) transforms business, economy and society by transforming experiences and relationships between stakeholders and citizens (Loureiro et al. 2021). The development of AI stimulates interest in studying cognitive and mental models of the human mind (Sergievskii 2020). AI is designed to simulate human intelligence to support and even expand human capabilities (Otte 2019). AI is a hot topic in society as it seems to expand and question human cognitive abilities (Neubauer 2021). Artificial Intelligence is the science of how to make computers perform intelligent tasks that could only be performed by humans in the past (Huang et al. 2019). AI is a multidisciplinary technology that may concern the following: the integration of cognition, machine learning, emotion recognition, human–computer interaction, data storage and decision making (Lu 2019). AI not only requires logic of thinking and imitation; emotions are an indispensable part thereof (Zhang and Lu 2021). With all the differences in views on the creation of AI (Artificial General Intelligence, AGI), the main obstacle to the creation of AGI is the complexity of describing the real world (Malinetsky and Smolin 2021). Nowadays, numerous AI applications are at least partly powered by (deep) artificial neural networks (SSN). SSN are found in many different varieties, but they all consist of a collection of largely identical processing "units" hierarchically ordered, just as biological neurons form biological neural pathways. In fact, psychology and neuroscience have had a strong influence on the development of the ANN network (van der Maas et al. 2021). Thus, AI de facto opens a new dimension, not only to the construction of railway systems, but above all to thinking in the virtual space created for new platforms in the technology-digitalization-human–robot relationship. AI-based breakthrough technologies and those using innovation in generating, transmitting and processing high-precision "straight from the track" data have a significant impact on increasing the efficiency, safety and attractiveness of rail transport. Both passenger and freight railways are already preparing for the introduction of fully autonomous trains. At the moment, trains without a train driver operate only on separate lines, and autonomous control systems are used in the subway, for example in Paris, Sydney or Shanghai. These solutions are developed thanks to artificial intelligence, which nowadays is used in high-speed rail systems around the world. Extremely strict safety requirements apply for rail transport. There is a certain reluctance to replace humans who make decisions in critical situations. Legal standards also pose barriers to the development of artificial intelligence in railways: according to the generally applicable law, a railway vehicle must have at least one train driver, which limits the development of autonomous vehicles based on artificial intelligence. However, AI will certainly be widely implemented as an innovative solution which increases the efficiency of the rail transport system in the near future after discussion among stakeholders interested in the development and operation of the railway system and overcoming railway safety barriers.

8.4 Simulations, Computer Games, Videos and Virtual Reality (VR) and Railway Traffic Safety Management

Digitalization, as a continuous process of convergence of the real and virtual world, which is moving towards cyber-physical systems, is now a major accelerator of innovation and dynamic changes in rail transport. It is therefore vital to be aware that not all solutions and their direct accommodation are possible. At the forefront is the possibility of using various types of virtual reality simulations. It is already clear that suitable modern tools that create virtual reality in different types of simulations should be used for the implementation of periodic instructions for people who are critical to railway safety. For this purpose, train driving simulators, simulators for train dispatchers and level crossing attendants and other such positions should be used systematically, together with the use of various types of computer games and educational films. In the course of the development of railway simulators, depending on the intended purpose and the group of recipients, certain types of railway simulators have evolved, which can be classified as follows:

a. full-scale (cabin) simulators: with a motion mechanism; without a motion mechanism;
b. desktop simulators;
c. compact simulators: stationary; mobile (e.g. in special trailers);
d. desktop simulators on PCs (low-cost simulators).

Full-scale (cabin) simulators are equipped with a fully closed training station, imitating the interior of an actual traction vehicle. The person who undergoes the simulation session is visually and acoustically isolated from the environment, which significantly improves the level of realism of the simulation. The image, depending on need, can be: displayed on monitors which imitate the view through the windows of the vehicle; projected on a flat or spherical/cylindrical external screen located outside the simulation cabin; or displayed directly on cabin windows (on-screen technology). The cabin can be equipped with a motion mechanism, consisting of a special platform on actuators (electric actuators are predominant at this point in time), illustrating movement with three or six degrees of freedom. It is understood that, for the safety of the trainee, only one training position should be available per instructor. Full-scale simulators are gaining recognition, for example, in Germany. In general, simulators also allow for the mapping of the system of generating an image of the railway line in front of the train, together with a system of visualization of the railway line, its immediate surroundings, various signals and images on indicators, as well as the situation on the track. Simulators include sound-generating systems which imitates hearing sensations occurring when working in the cabin of an actual traction vehicle; a system for generating the mechanics of train movement affecting both the visualization system (speed of approaching the picture of the railway line on the screen) and the mechanism of movement of the train driver's seat (jerking upon start and braking, vibrations while driving, bouncing of the traction vehicle on switches and bumps) (Brona et al. 2015).

The use of railway simulators is developing very dynamically. Their development is followed by provisions in legal regulations, for example, concerning the conduct of courses and the examination of staff performing tasks which are critical to railway traffic safety. Simulators based on digital solutions test non-standard critical situations and thus improve operators' ability to respond to unusual or rare situations during the normal operation of railway systems. They allow for the mitigation of the risk of human errors in rail traffic operation. It seems reasonable to note that in the near future, railway simulators in an exceptionally large number of railway areas will dominate the processes of training railway staff in order to improve their practical interdisciplinary skills.

8.5 The Internet of Things and Interface Management in the Process of Improving Railway Traffic Safety

The Internet of Things has created a global network supporting ubiquitous computing (Bandyopadhyay et al. 2011). The technological ecosystem underlying IoT is mainly the integration of cyber-physical systems (CPS) (Akkaya et al. 2016). It also includes cloud computing, automation (e.g. smart robots on product assembly lines), the Internet of Services, wireless technologies, augmented reality and coaxial processing, and more besides. In this cognitive space, IoT represents futuristic processing, communication and technology (Goyal et al. 2021). This is particularly true for rail value chains. Supply Chain Management (SCM) is largely based on the use of well-analyzed data, simply because data-based decisions lead to better results in complex business environments (Speranza 2018). IoT perceptual capabilities are based on a variety of identification and tracking technologies that allow for the remote monitoring of physical objects without having to be within sight (Atzori et al. 2010). It is important to develop specialist software using IoT for specific conditions of rail transport. For example, an intelligent rail logistics identification system using IoT can be used in real time for the effective control and management of various aspects of rail logistics. The intelligent identification of the rail logistics system using IoT is a complete system with a combination of hardware and software using the computer as the core (Guo et al. 2012).

Based on the Internet of Things, a rail vehicle planning system can be developed. This system is designed to quickly and efficiently manage relevant information, reduce human, material and financial resources, and ensure the security and accuracy of relevant information while reducing the workload (Zhang et al. 2021). The Internet of Things also applies to railway traffic control systems. The future includes track-side equipment equipped with standard IP-based interfaces and digital interlocking lever frames that are able to control railway traffic across the region. Digitalization allows for the introduction of modular solutions, independent of the manufacturer, and thus more favorable in terms of operation and maintenance as well as future modernizations for the infrastructure manager. In its interdisciplinarity, the Internet of Things expands the range of solutions based on this concept. As part of this option, the following ideas can be included: the Internet of Computers (IoC), the Internet of Devices (IoD), the Internet of Services (IoS), the Internet of Things (IoT), the Industrial Internet of Things (IIoT), the Internet of Everything (IoE), the Internet of Things, Services and People (IoTSP), the Internet of Robots (IoRT), and the Internet of Emotional Things (IoET). The Internet of Things in the rail transport sector may first apply in the following areas: Infrastructure Monitoring Systems, Real Time Train Tracking System, the interaction of ATO systems with ETCS (European Train Control System), and other applications. The Internet of Things (IoT)—i.e. connected devices collecting and exchanging data—is a huge opportunity to increase cost efficiency and improve safety across the railway system. Interfaces between devices and the creation of coherent systems mean that smart communication applies not only to people but also systems, which is particularly important in high-complex rail technical systems. This idea fits into the theory of intelligent systems of systems. This means that in the near future the communication potential of railway system equipment can increase the level of excellence of railway systems to a much higher level than is currently the case.

8.6 Predictive Maintenance of Railway Infrastructure and Railway Vehicles in Terms of Safety

The maintenance of railway infrastructure as well as railway vehicles is one of the crucial aspects of railway traffic safety management. In both the theory and practice of technical equipment maintenance management, two approaches are applied, namely the preventive and predictive approaches. The preventive approach consists of conducting regular and planned inspections aimed at detecting and preventing failures. The intervals at which they are made are determined according to the specificity of the equipment and its failure rate—inspections are usually conducted every few weeks, months or years. For this purpose, appropriate maintenance system documentation is developed for each type of equipment or railway vehicle. In all respects, the manufacturer's recommendations and operational experience must be complied with. The preventive maintenance of modern and run-down rail infrastructure alike should be part of the digital transformation process in the era of digitalization. Increasingly, infrastructure managers and rail operators have already implemented asset management systems which take the elements of resource identification and the need to supervise them according to specific rules and principles into account.

According to the predictive strategy, action is taken at the most appropriate time—that is, just after the first signs of an impending failure appear, but before it occurs. This involves continuous monitoring of the condition of the device to immediately detect possible disturbing symptoms. Then intervention occurs to prevent the problem from developing and, as a consequence, from failure. If the measurement data is archived, the more this is done over time, the fuller the knowledge of the device is, which may prove useful in the future for it or other similar devices. Predictive and responsive maintenance are among the corrective methods used. This category includes action taken to restore the proper functioning of the device after a failure or to record the deterioration of the work ultimately leading to it. The use of sensors is the basic solution of modern predictive maintenance systems. They are built into many places on the map of the rail transport system, creating intelligent solutions based on the use of Big Data and the detectable state of operation of technical systems. The condition for the use of predictive maintenance solutions is the use of solutions from the area of the digital economy; it is not possible to implement these processes in an analog manner. Digital solutions support technical department management. They facilitate the planning and control of maintenance work, affect the efficiency of maintenance services and reduce costs. Automatic notifications of failures significantly reduce the response time to incidents, and the data recorded in the system provide a basis for estimating the most important indicators such as MTBF, MTTR, and MTTF. Recording activities on mobile devices and generating electronic documents eliminates most possible mistakes, and helps managers delegate responsibilities and report the work of the entire department, as well as analyzing and monitoring the effectiveness of machines/devices. This has a very strong impact on preventing defects, failures, and consequently rail accidents and disasters. An example of the course of the predictive maintenance process of railway infrastructure is presented in Fig. 8.1.

The presented scheme of the predictive maintenance process of railway infrastructure is based on an IT solution combining the use of mobile devices for maintenance staff, the option of self-remediation of faults by the diagnostic system and the installation of sensors in various locations of the device or railway system. It is based on intelligent solutions supported by a knowledge base. An extensive system of identification and traceability of the predictive maintenance system allows for the detection of emergency states—weak signals before a fault occurs. A rapid response allows for the mitigation of the risk of faults, failures, and in the worst case, disasters. Such functionalities of maintenance systems ensure that the operational readiness and safety indicators are at the expected level. They are now essential to the management and monitoring of the safety of railway systems.

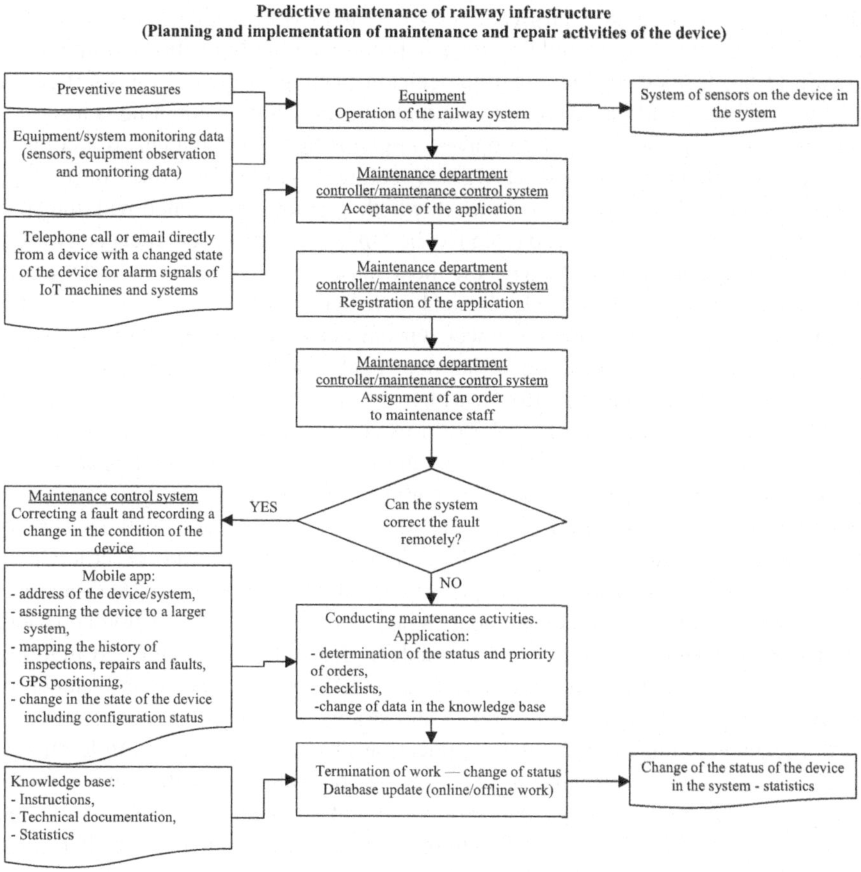

Fig. 8.1 Example of the predictive maintenance process of railway infrastructure (*Source* Own study)

8.7 Big Data and its assumptions in shaping digital solutions in rail transport

The development of database systems, in terms of the concepts of both Big Data and Open Data, creates the conditions for innovative solutions in the field of rail transport. Collecting large amounts of data, processing and using it to meet unrecognized functionalities is an area that has not yet created the opportunity to build optimal solutions. Open databases offer this opportunity and reveal new unconventional ways of creating value for railway system operators and users, including with regard to safety objectives in rail transport. Predictive analytics is used to predict future scenarios of events, and prescriptive analytics provides knowledge of how to reduce these events by modeling responses to the risks of their occurrence. Digital data engineering provides railway market operators with extensive knowledge of events and behaviors in the railway system, preferences and intentions of operators.

Big Data for analysis and dynamic decision making in the rail transport sector is supported by cloud-based solutions in many cases. Cloud computing is a service model for sharing and using IT resources, implemented through a network connection and accounted for the use of a given feature. The cloud is based on making resources available to the user on request, which is usually associated with the "pay as you go" financial model, i.e. paying for the functions, data or server capacity used. Infrastructure-as-a-Service (IaaS) is the most "basic" cloud service, whereby a user purchases a certain amount of IT, physical or virtual infrastructure from a provider. It can be disk space, computing power, virtual network area (VLANs) or even a pool of IP addresses. Platform-as-a-Service (PaaS) means access via a browser to a ready-made development platform, which the client can modify and configure to satisfy their business needs. The platform usually includes an operating system, a programming language environment, a database and a web server. With PaaS, developers can create and test their own cloud applications without having to purchase a complex and costly license for a given programming environment and the necessary hardware. This is an immensely popular solution in the rail transport sector. Cloud computing facilitates the use of concepts which enables the services of numerous entities, namely multi-tenancy management. When it comes to enabling a cloud provider to take full advantage of the scalability and flexibility of the cloud, multi-availability is a key feature of the cloud that allows providers to share the same instance of service, computing, access to resources and storage among different operators. In multi-tenancy architecture, a single application instance supports multiple clients. Every client is called a tenant. In this way, the application in multi-tenancy architecture can share a dedicated instance of configuration, data, user management, and other properties. Multi-tenancy refers to the architectural principle in which a single software instance works on SaaS servers, organizations (tenants) serving multiple clients. Multiple tenants contrast with multiple architectural instances where separate instances of software (or hardware and software)/systems are created for different customer organizations (Banks et al. 2009). Multiple tenants are a new design for software architecture with one instance of application (or data customization and configuration) that works in the service provider's infrastructure (Ru et al. 2021). Figure 8.2 presents a solution for the architecture of complex data management systems according to the concept of multi-tenancy management used in the railway industry.

Multi-tenant applications and microservices fulfill the combined requirements of multiple user groups, departments or companies (i.e. tenants). By allowing tenants to share the same instances, the number of required instances, as well as the cost of developing, maintaining and implementing such microservices, can be reduced (Furda et al. 2018). The multi-tenant component architecture is shown in Fig. 8.3.

The application space of the multi-tenant architecture concept is therefore very wide and is especially useful for complex systems where many entities operate and data is exchanged on many layers and a significant number of thematic areas in a dynamic time period (Kabbedijk et al. 2015; Zeng and Plale 2016; Bernabe et al.

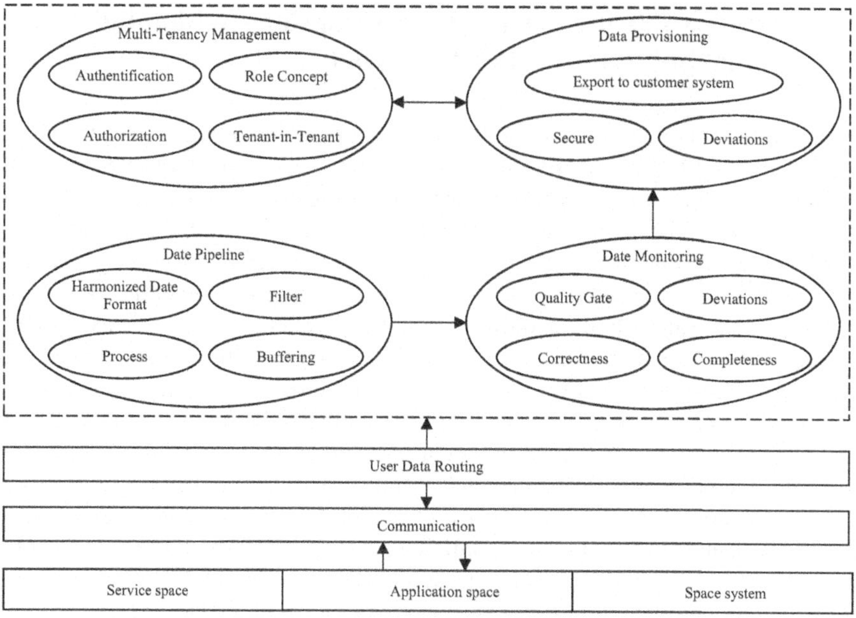

Fig. 8.2 Architecture of complex data management systems based on the concept of multi-tenancy management. (*Source* Own study based on Ci4Rail, Computing Intelligence for Rail & Public Transport. EdgeFarm. The Cloud-based IoT platform for holistic fleet-wide management of edge devices, distributed applications and user data)

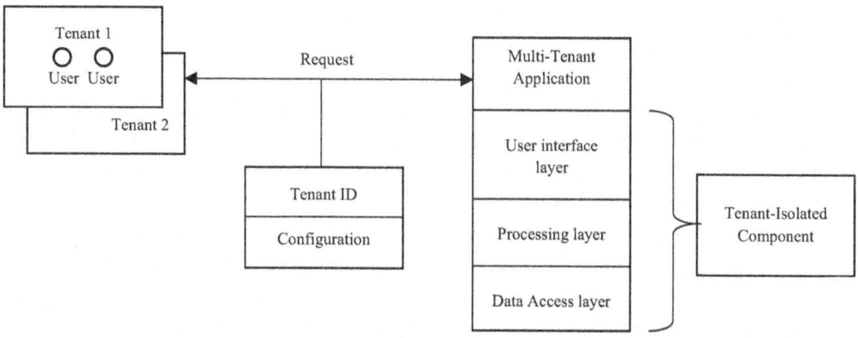

Fig. 8.3 Multi-tenant component architecture. (*Source* Own study based on Furda A, Fidge C, Barros A (2018) A practical approach for detecting multi-tenancy data interference. Sci Comput Program 163: 160–173, 16, 10.1016/j.scico.2018.04.006)

2014; Kuppusamy et al. 2015). This model of data management in rail transport systems allows for the achievement of a range of results which are conducive to achieving a high level of the optimization of transport processes and information flow.

8.8 Industry 4.0 and 5.0 and its Assumptions in Shaping Digital Solutions in Rail Transport

The term Industry 4.0, the fourth industrial revolution, has emerged since the beginning of the 2020s. The label Industry 4.0 has been given to smart industry in the twenty-first century in the context of earlier industrial developments since the industrial revolution. The main essence of Industry 4.0 is not a new and groundbreaking technological invention that will mean that parts are made faster in the production process. A profound system change in the production process is important. Its primary task is to streamline, clarify and further link the production process. This systemic change should be a kind of fusion of classical industry and digital technologies over the Internet. The main changes which the Industry 4.0 concept is built on include a communication environment, environmental identification, extensive data collection and non-standard decisions in different areas (Gerhátová et al. 2021). Industry 4.0 describes the growing digitalization and automation of the production environment and the creation of digital value chains for communication between products, their environment and business partners (Dallasega et al. 2018). It should be noted that the core attributes of Industry 4.0 include decentralization, interoperability, real time, security, service orientation, modularity and visualization (Habib ur Rehman et al. 2019). The Industry 4.0 standard revolutionized the manufacturing sector by integrating several technologies such as artificial intelligence (AI), the Internet of Things (IoT), cloud computing, cyber-physical systems (CPS) and cognitive processing. The main principle of Industry 4.0 is to make the manufacturing industry "smart" by interconnecting machines and devices that can control each other throughout the life cycle (Xu 2020). Industry 4.0 focuses on improving mass productivity and efficiency by providing intelligence between machine learning (ML) devices and applications (de la Peña Zarzuelo et al. 2020).

The concept of Industry 5.0, which undoubtedly already applies to rail transport, is also beginning to be used in the economy. Industry 5.0 is considered to be another industrial evolution; its aim is to exploit human creativity and expertise in collaboration with efficient, intelligent and accurate machines, so as to achieve resource-efficient and user-friendly manufacturing solutions compared to Industry 4.0. These are numerous and promising technologies. Applications are expected to support Industry 5.0 to increase the production and delivery of custom products spontaneously (Maddikunta et al. 2021). Industry 5.0 promotes more skilled jobs compared to Industry 4.0 because so-called white-collar workers work with machines. The focus on Industry 5.0 primarily involves mass personalization, which will be conducted by people and robots. In Industry 4.0, robots are already actively involved in large-scale production, while Industry 5.0 aims primarily to improve consumer satisfaction. Industry 4.0 focuses on CPS connectivity, while Industry 5.0 connects to Industry 4.0 applications and establishes a link between collaborative robots (cobots). Another interesting advantage of Industry 5.0 is to provide more ecological solutions compared to existing industrial transformations, none of which are focused on environmental protection (Demir et al. 2019). Two visions of Industry

5.0 are emerging. The first is "man-robot cooperation". In this vision, a robot and people will cooperate whenever and wherever possible. People will focus on tasks that require creativity and robots will do the rest. Another vision of Industry 5.0 is the bioeconomy (Demir and Halil 2018). Industry 5.0 is the age of the Social Smart factory, where cobots communicate with people (Koch et al. 2017). Industry 5.0 forces various industry practitioners, IT professionals and philosophers to focus on considering human factors in industrial systems technologies (Friedman and Hendry 2019). It is crucial to refer to the main issues related to Industry 5.0 at the level of the organization. Issues related to the integration of robots into organizations include:

- Evolution of organizational behavior,
- Acceptance of robots in the workplace,
- Evolution of organizational structures and workflows,
- Evolution in work ethics,
- Discrimination against robots or people,
- Privacy and trust in a common human and robot working environment,
- Education and training,
- Redesigning workplaces for robots (Demir et al. 2017).

It should be noted that, according to one definition, a robot is a machine that senses, thinks and works (Bekey 2005). The Oxford dictionary defines a robot as a machine that can perform a complicated series of tasks automatically, especially one programmable by a computer. Webster's dictionary defines a robot as a machine that resembles a living creature and performs various complex actions (such as walking or speaking) of a human being. It is important that legal and regulatory issues pertaining to the use of artificial intelligence, robots, drones and cyborgs are currently subject to a great deal of scientific research (Demir et al. 2019). Thus, the concept of responsible innovation has arisen (Hankins 2018). In the context of scientific discussion about the rail transport sector in terms of safety, one should refer to the fact that, for example, Sukmono and Junaedi proposed to use the Fifth Industrial Revolution in the context of disaster management (Sukmono and Junaedi 2020). Therefore, the application of the concept of Industry 5.0 in rail transport is diverse and can be a platform for technological changes in many areas of innovation in rail transport.

8.9 A Digital Twin as a Concept of Improving Railway Traffic Safety

A digital twin is a multi-physical, multi-scale and integrated probabilistic simulation of a complex product using the best available physical models, sensor updates, etc. to reflect the life of the corresponding twin. These rather general definitions do not focus on the most interesting properties of a DT: real-time data acquisition (Glaessgen and Stargel 2012).

The concepts of the digital model, the digital shadow and the digital twin were introduced by Werner K, based on data exchange flows, depending on the degree of data and their integration.

Digital model: A digital model that does not use any form of automatic data exchange between physical and digital objects.

Digital Shadow: Based on the definition of a digital model, there is an automated one-way flow of data between the state of the existing physical object and the digital object where applicable. This is already possible at the construction stage.

Digital Twin: If further data flows between existing physical and digital objects and are fully integrated in both directions, this could be called a digital twin. In such a connection, the digital object can also perform the function of controlling the access to the physical object. There may also be other physical or digital objects that cause changes in the state of a digital object. The change in the state of the physical object leads directly to the change in the state of the digital object and vice versa (General Electric). A digital twin is a simulation of the physical object itself to predict future system states (Gabor et al. 2016). It is also a digital representation of a real-world object with an emphasis on the object itself (Canedo 2016). A digital twin is also a virtual substitute for real-world objects consisting of virtual representations and communication capabilities consisting of intelligent objects acting as intelligent nodes in the Internet of Things and Services (Schluse and Rossmann 2016). It is also an integrated multi-physical, multi-scale, probabilistic simulation of the as-built system, enabled by Digital Thread, which uses the best available models, sensor information and inputs to reflect and predict activity/results over the lifetime of the corresponding physical twin (Kraft 2016). It is also a digital product as equivalent to a physical product (Rios et al. 2015).

A digital twin is also a virtual representation of a real product in the context of cyber-physical systems (Schroeder et al. 2016). In terms of the applicability of this concept, it can, inter alia, be used widely in rail transport in railway traffic control systems. The determinant of the development of this concept is the direct relationship in the area of data and information in the real system space and virtual mapping space system. Thus, the concept of the digital twin of the railway traffic control system becomes important. Then a digital representation of the actual materializing structure is created as a model of process architecture through the set simulation models. A digital twin is equipped with two-way data transmission from and to the real system, while being able to describe the current state and save archive data. Crucial factors which facilitate the use of the digital twin are its conceptualization elements, the ability to synchronize (active or non-active synchronization), interface layouts and application opportunities. As regards interface construction, these are mainly two types of interfaces: machine-human and machine-machine. In relation to its structure, input data (raw or processed), data links and creation time are important. It is interesting to determine the priority in the emergence of a digital twin in terms of the time of creation. Did the initiation come from a physical part, a digital part, or both at the same time? The accuracy of the mapping (identicalness or partial accuracy) of the model, as well as the methods of communication between the digital twin and its actual counterpart in a one-way and two-way perspective, also come to the fore. The use of data received from the actual system is related to the logic of data processing. Dynamic processes in a digital twin are forecasting and optimization processes. It is also worth noting that a digital twin can function in various forms. It usually occurs as part of a real system or as a separate entity. As a consequence, for modern railway traffic control systems, in terms of the design of ERTMS/ETCS systems, the digital representation of the ETCS architecture model and the train simulation model is shaped using bidirectional data transmission to and from the real system together with aggregated data including current status and archival data. It should be noted that the ERTMS/ETCS digital twin has an application capacity throughout its life cycle (concept, design, commissioning, verification, acceptance, certification, maintenance, modification, and decommissioning).

8.10 Digital Solutions in Rail Transport Taking into Account the Ecological Criteria Used

One of the crucial areas of rail transport development is the greening thereof by means of digital solutions. This is based on a specific triad comprising the electrification, digitalization and decentralization of the railway system. In this case, the greening of the railway energy supply chain is an irreversible trend, and photovoltaics (PV) provides the most suitable type of renewable energy for integration into railways. Generating photovoltaic energy with dynamic loads on the railways increases the flexibility needed to maintain the balance of load generation. The new evolutionary model of the Rail Power Supply System (RESS) for railway photovoltaic integration systems (RPISs) proposes the construction of a three-in-one integrated station (TSIIS) type of 'information traction-storage' (Ning et al. 2021). The use of a green supply chain in the railway system opens the way towards a new dimension of rail traffic. The elimination of excessive CO^2 emission in rail transport is an important development objective and should be supported by dedicated digital solutions. It should also be pointed out that, due to the fact that train traffic is subject to an accurate timetable, rail transport offers the possibility of the effective application of model methods that take into account the dynamics of the traffic and allow for an accurate calculation of energy demand and an assessment of CO^2 emission, which should now be strongly supported by digital solutions. Such solutions include recuperation, i.e. energy recovery during vehicle braking, which is one of the key tools for optimizing energy consumption on the railway. Depending on the specific route, recuperation can save 20–25% of energy. This concept makes it possible to settle the energy generated by the process of recuperation during braking of traction vehicles and to reduce the operating costs of rolling stock. However, the use of recuperation does not complete the list of possible energy-saving measures on the railways. Equipping vehicles with energy monitoring devices allows for the implementation of eco-driving principles—eco-driving train drivers who, like drivers, can reduce their energy consumption through their driving style. Eco-driving offers savings of up to 5–10% of energy consumed. The digital economy solutions are therefore again applicable to this purpose. This widespread and broad approach to the digitalization of railways transforms this system into a fully digitalized railway business model.

8.11 Digital Solutions in Rail Transport for Communication Systems in Railway Traffic Control

The most commonly used railway mobile radio system is the GSM-R system, which is a Class A unified digital radio communications system designed to provide voice communication and digital data transmission necessary for the operation of ETCS level 2. The GSM-R system together with ETCS form the European Rail Traffic Management System (ERTMS). The GSM-R system standard is defined by EIRENE specifications, which form a set of functional and system requirements. GSM-R is a duplex cellular structure designed exclusively for railway purposes. The architecture of the GSM-R system is based on the ETSI GSM Phase 2 + standard. The GSM-R system works in a dedicated GSM-R 900 MHz band. The GSM-R network is a closed network, i.e. the possibility of making voice calls and sending and receiving SMS messages is restricted to users of GSM-R domestic or foreign networks, except for voice calls to public emergency numbers. The GSM-R network implements:

1. Voice communication services: point-to-point voice calls, emergency calls, broadcast calls, group voice calls, and conference calls.
2. Voice communication-related services: call priority, Advanced Speech Call Items (ASCI) such as: redirecting calls, suspending calls, etc., automatically answering calls, and blocking incoming and outgoing calls.
3. Specific railway services: functional numbering, location-dependent addressing, and REC emergency calls.
4. Short Message Service (SMS).
5. Data transmission service: for ETCS level 2, GPRS standard.

The FRMCS concept is emerging in the context of the development of new mobile radio solutions. FRMCS is designed to become a global standard of rail transport. This mobile broadband technology helps improve safety and operational efficiency, support innovative passenger services and accelerate the digital transformation, all of which will help compete for travelers. FRMCS also minimizes network latency and uses cloud technology to help automate train operation and support M2M broadband communication. FRMCS is a communication standard which covers demanding mobile radio systems. FRMCS will act as a catalyst for digital railways, making future services even more reliable, and will increase the capacity of existing railway networks and optimize system costs. FRMCS plays a crucial role not only by replacing the existing GSM-R communication system, but also by enabling the deployment of a wide range of new applications such as automatic train operation, the Internet of Things (IoT) and efficient maintenance. This system includes the standardization of the architectural frameworks and interfaces of the on-board system, taking into account potential long and diverse migration and life cycle management schedules. 5G can help to take full advantage of FRMCS migration. It offers high capacity, efficiency and reliability, as well as support for huge MTC and IoT technologies. With 5G-based FRMCS, extremely reliable and critical communication with low latency can be handled. A platform for innovative applications can also be built—from group video calls and automated train control to predictive maintenance and high-speed on-board Wi-Fi, which will reduce operating costs and improve passenger experience. Future X architecture paves the way to 5G and FRMCS, ensuring a smart and dynamic communication platform that supports all railway systems, processes and activities. This will allow for the creation of new performance and support for new applications, ensuring safe and reliable communication for people, sensors, trains, video monitors and automated train control systems (Nokia). FRMCS will be based on 3GPP building blocks, and gaps will be covered by ETSI specifications.

1. The system will have an FRS and an SRS, to provide the necessary means to operate the FRMCS networks and to guarantee their interoperability in the context of cross-border operations. The FRS and the SRS will complete the FRMCS system and will be the base in the CCS TSI for the Train-Radio and ETCS critical applications (Mandoc 2019).
2. The radio communication system currently used in rail operations, namely the railway mobile radio (GSM-R installation), is based on specifications that were finalized 20 years ago, and as they are technologically outdated, it is unlikely that industry will provide support for GSM-R installations for a long period after 2030. The Future Railway Mobile Communication System (FRMCS) will replace the GSM-R installation and will become one of the essential elements of the European Rail Traffic Management System (ERTMS). FRMCS will support railway digitalization and service innovation. The installation of GSM-R and its successor(s), including FRMCS, is referred to as Railway Mobile Radio (RMR).
3. Compared to the GSM-R, FRMCS offers a higher quality of services, uses the spectrum more efficiently and is more cost-effective. It is also planned for the system to provide more application possibilities such as Automatic Train Operation (ATO) or a train driver support system such as the Connected Driver Advisory System (C-DAS). Further applications are expected to be introduced. Railway-critical FRMCS applications such as critical infrastructure monitoring and control can be operated efficiently by means of Narrowband IoT technology (NB-IoT). FRMCS should be able to take new applications and technological developments into account over a longer period of time, as railway communication systems have a significantly longer life cycle compared to public electronic communications networks and services.
4. It is therefore necessary to harmonize the frequency bands to allow for the introduction of FRMCS.

5. Access to a sufficiently harmonized spectrum for RMR is essential to allow for the parallel operation of the GSM-R system and its successor during the approximately 10-year migration phase from GSM-R to FRMCS and the use of new critical applications for railways during the migration period and beyond (Commission Implementing Decision (EU) 2021/1730). GSM-based systems are crucial for the digital transformation of the rail transport sector. They provide support for train control systems and ensure safe communication and data transmission.

8.12 Conclusion

Digital technologies are accelerating progress in the rail transport sector. The multiplicity of their applications creates the need to set new safety standards. Technological progress is being created in a large number of areas and levels of operation of railway enterprises. Big Data sets in particular allow for new possibilities in the field of safety, diagnostics of equipment and railway vehicles, and analytics in various areas of operation of railway systems. Data management through its collection, analysis and use allows for the performance of thus far unfamiliar functionalities of railway systems. The implementation of solutions based on Big Data enriches the configuration of digital business models of railway enterprises and the railway system. The use of sensors and hardware, simulations, artificial intelligence concepts, digital twins, the Internet of Things concept, Industry 4.0 and 5.0, as well as many other solutions, has completely changed the architecture of and approach to railway systems. This change is positive, as in addition to improving safety, it creates new opportunities to improve the quality and attractiveness of inter-modal rail transport.

References

Akkaya I, Derler P, Emoto S, Lee EA (2016) Systems engineering for industrial cyber–physical systems using aspects. Proc IEEE 104(5):997–1012. https://doi.org/10.1109/JPROC.2015.2512265

Atzori L, Iera A, Morabito G (2010) The Internet of Things: a survey. Comput Netw 54(15):2787–2805. https://doi.org/10.1016/j.comnet.2010.05.010

Bandyopadhyay S, Sengupta M, Maiti S, Dutta S (2011) Role of middleware for Internet of Things: A study. Int J Comput Sci & Eng Surv 2(3):94–105. https://doi.org/10.5121/ijcses.2011.2307

Banks D, Erickson J, Rhodes M (2009) Multi-tenancy in cloud-based collaboration services. Technical Report HPL-2009–17, HP Laboratories Palo Alto

Bekey GA (2005) Autonomous robots: from biological inspiration to implementation and control. The MIT Press, Cambridge, USA

Bernabe JB, Perez JMM, Alcaraz Calero JM, Garcia Clemente FB, Martinez Perez G, Gomez Skarmeta AF (2014) Semantic-aware multi-tenancy authorization system for cloud architectures. Futur Gener Comput Syst 32: 154–167, https://doi.org/10.1016/j.future.2012.05.011

Borges AFS, Laurindo FJB, Spínola MM, Gonçalves RF, Mattos CA (2021) The strategic use of artificial intelligence in the digital era: Systematic literature review and future research directions. Int J Inf Manag 57: 102225, https://doi.org/10.1016/j.ijinfomgt.2020.102225

Brona P, Dąbrowski A, Koszela J, Król D, Szafrański Z, Wantoch-Rekowski R (2015) Projekt nowoczesnego demonstratora symulatora dla operatorów pojazdów szynowych. Technika Transportu Kolejowego 1–2:41–49

Canedo A (2016) Industrial IoT lifecycle via digital twins. In: CODES '16: proceedings of the eleventh IEEE/ACM/IFIP international conference on hardware/software codesign and system synthesis, 29, https://doi.org/10.1145/2968456.2974007

Ci4Rail, Computing Intelligence for Rail & Public Transport. EdgeFarm. The Cloud-based IoT platform for holistic fleet-wide management of edge devices, distributes applications and user data

Commission Implementing Decision (EU) 2021/1730 of 28 September 2021 on the harmonized use of paired frequency bands 874.4–880.0 MHz and 919.4–925.0 MHz and the unpaired frequency band 1 900–1 910 MHz for the Railwy Mobile Radio (notified under document C(2021) 6862)

Dąbrowa-Bajon M (2007) *Podstawy sterowania ruchem kolejowym. Funkcje, wymagania, zarys techniki.* Warszawa, Oficyna Wydawnicza Politechniki Warszawskiej

Dallasega P, Rauch E, Linder C (2018) Industry 4.0 as an enabler of proximity for construction supply chains: a systematic literature review. Comput Ind 99:205–225. https://doi.org/10.1016/j.compind.2018.03.039

de la Peña Zarzuelo I, Soeane MJF, & Bermúdez BL (2020) Industry 4.0 in the port and maritime industry: a literature review. Journal of Industrial Information Integration 20: 100173, https://doi.org/10.1016/j.jii.2020.100173

Demir KA, Cicibaş H (2018) The next industrial revolution: industry 5.0 and discussions on industry 4.0. In: Industry 4.0 From the Management Information Systems Perspectives. Peter Lang Publishing House, pp 247–260

Demir KA, Caymaz E, Elci M (2017) Issues in integrating robots into organizations. The 12th International Scientific Conference on Defense Resources Management in the 21st Century, November 09–10, Brasov, Romania

Demir KA, Döven G, Sezen B (2019) Industry 5.0 and human-robot co-working. Procedia Computer Science 158:688–695. https://doi.org/10.1016/j.procs.2019.09.104

Friedman B, Hendry DG (2019) Value sensitive design: shaping technology with moral imagination. The MIT Press, Cambridge

Furda A, Fidge C, Barros A (2018) A practical approach for detecting multi-tenancy data interference. Sci Comput Program 163:160–173. https://doi.org/10.1016/j.scico.2018.04.006

Gabor T, Belzner L, Kiermeier M, Beck MT, Neitz A (2016) A simulation-based architecture for smart cyber-physical systems. In: 2016 IEEE International Conference on Autonomic Computing (ICAC), 374–379, https://doi.org/10.1109/ICAC.2016.29

General Electric, *Digital Twin*, www.ge.com/digital/applications/digital-twin [accessed 10 Nov 2021].

Gerhátová Z, Zitrický V, Klapita V (2021) Industry 4.0 implementation options in railway transport. Transportation Research Procedia 53:23–30. https://doi.org/10.1016/j.trpro.2021.02.003

Glaessgen EH, Stargel DS (2012) The digital twin paradigm for future NASA and US air force vehicles. Paper for the 53rd Structures, Structural Dynamics, and Materials Conference: Special Session on the Digital Twin

Goyal P, Sahoo AK, Sharma TK (2021) Internet of things: architecture and enabling technologies. Materials Today: Proceedings 34(3):719–735. https://doi.org/10.1016/j.matpr.2020.04.678

Gradowski P (2003) Aplikacje Europejskiego Systemu Zarządzania ruchem kolejowym (ERTMS) w zarządach kolejowych. Problemy Kolejnictwa 137(138):85–106

Guo Z, Zhang Z, Li W (2012) Establishment of intelligent identification management platform in railway logistics system by means of the Internet of Things. Procedia Eng 29:726–730. https://doi.org/10.1016/j.proeng.2012.01.031

Habib ur Rehman M, Yaqoob I, Salah K, Imran M, Jayaraman PP, Perera C (2019) The role of big data analytics in industrial Internet of Things. Futur Gener Comput Syst 99: 247–259, https://doi.org/10.1016/j.future.2019.04.020

Hankins J (2018) Responsible innovation, a narrative approach. Bergamo, Università degli Studi di Bergamo, (Collana della Scuola di Alta Formazione Dottorale; 9), https://doi.org/10.6092/978-88-940721-3-6

Huang B, Huan Y, Xu L, Zheng L, Zou Z (2019) Automated trading systems statistical and machine learning methods and hardware implementation: a survey. Enterp Inf Syst 13(1):132–144. https://doi.org/10.1080/17517575.2018.1493145

Kabbedijk J, Bezemer CP, Jansen S, Zaidman A (2015) Defining multi-tenancy: a systematic mapping study on the academic and the industrial perspective. J Syst Softw 100:139–148. https://doi.org/10.1016/j.jss.2014.10.034

Koch PJ, van Amstel MK, Debska P, Thormann MA, Tetzlaff AJ, Bøgh S, Chrysostomou D (2017) A skill-based robot co-worker for industrial maintenance tasks. Procedia Manuf 11:83–90

Kraft EM (2016) The US air force digital thread/digital twin—life cycle integration and use of computational and experimental knowledge. The evolution of integrated computational/experimental fluid dynamics, In: 54th AIAA Aerospace Sciences Meeting, pp 1–22. https://doi.org/10.2514/6.2016-0897

Kuppusamy S, Kaniappan V, Thirupathi D (2015) Design and development of multi-tenant web framework. Procedia Comput Sci 48:180–191. https://doi.org/10.1016/j.procs.2015.04.168

Loureiro SMC, Guerreiro J, Tussyadiah I (2021) Artificial intelligence in business: state of the art and future research agenda. J Bus Res 129:911–926. https://doi.org/10.1016/j.jbusres.2020.11.001

Lu Y (2019) Artificial intelligence: a survey on evolution, models, applications and future trends. J Manag Anal 6(1):1–29. https://doi.org/10.1080/23270012.2019.1570365

Maddikunta PKR, Pham Q-V, Prabadevi B, Deepa N, Dev K, Gadekallu TR, Ruby R, Liyanage M (2021) Industry 5.0: A survey on enabling technologies and potential applications. J Ind Inf Integr: 100257. https://doi.org/10.1016/j.jii.2021.100257

Malinetsky G, Smolin V (2021) The artificial intelligence influence on real sociality. Procedia Comput Sci 186:344–351. https://doi.org/10.1016/j.procs.2021.04.155

Mandoc D (2019) FRMCS definition, specification and standardization activities. ERA CCRCC.

McCarthy J (1959) Programs with common sense at the wayback machine. In: Proceedings of the teddington conference on the mechanization of thought processes, 756–91. London, Her Majesty's Stationery Office.

Neubauer AC (2021) The future of intelligence research in the coming age of artificial intelligence—with a special consideration of the philosophical movements of trans—and posthumanism Intelligence 87: 101563, https://doi.org/10.1016/j.intell.2021.101563

Ning F, Ji L, Ma J, Jia L, Yu Z (2021) Research and analysis of a flexible integrated development model of railway system and photovoltaic in China. Renew Energy 175:853–867. https://doi.org/10.1016/j.renene.2021.04.119

Nokia. Future Railway Mobile Communication System (FRMCS). Ensure safe, dependable operations with FRMCS based on 5G. https://www.nokia.com/networks/solutions/frmcs/ [accessed: 10 Nov 2021]

Otte R (2019) Künstliche Intelligenz für Dummies [Artificial intelligence for dummies]. WILEY-VCH Verlag GmbH

Ríos J, Hernandez JC, Oliva M, Mas F (2015) Product avatar as digital counterpart of a physical individual product: literature review and implications in an aircraft. Transdiscipl Lifecycle Anal Syst 2:657–666. https://doi.org/10.3233/978-1-61499-544-9-657

Ru J, Yang Y, Grundy J, Keung J, Hao L (2021) A systematic review of scheduling approaches on multi-tenancy cloud platforms. Inf Softw Technol 132: 106478, https://doi.org/10.1016/j.infsof.2020.106478

Schluse M, Rossmann J (2016) From simulation to experimentable digital twins. In: 2016 IEEE International Symposium on Systems Engineering (ISSE), 1–6. https://doi.org/10.1109/SysEng.2016.7753162

Schroeder GN, Steinmetz C, Pereira CE, Espindola DB (2016) Digital twin data modeling with automationml and a communication methodology for data exchange. IFAC-PapersOnLine 49(30):12–17. https://doi.org/10.1016/j.ifacol.2016.11.115

Sergievskii V (2020) Super strong artificial intelligence and human mind. Procedia Comput Sci 169:458–460. https://doi.org/10.1016/j.procs.2020.02.225

Speranza MG (2018) Trends in transportation and logistics. Eur J Oper Res 264(3):830–836. https://doi.org/10.1016/j.ejor.2016.08.032

Subramaniam M, Iyer B, Venkatraman V (2019) Competing in digital ecosystems. Bus Horiz 62(1):83–94. https://doi.org/10.1016/j.bushor.2018.08.013

Sukmono FG, Junaedi F (2020) Towards industry 5.0 in disaster mitigation in Lombok island, Indonesia. Jurnal Studi Komunikasi (Indones J Commun Stud) 4 (3): 553–564. https://doi.org/10.25139/jsk.v4i3.2424

van der Maas HLJ, Snoek L, Stevenson CE (2021) How much intelligence is there in artificial intelligence? A 2020 update. Intelligence 87: 101548, https://doi.org/10.1016/j.intell.2021.101548

Xu LD (2020) Industry 4.0—frontiers of fourth industrial revolution. Syst Res Behav Sci 37 (4): 531–534. https://doi.org/10.1002/sres.2719

Zeng J, Plale B (2016) Argus: A multi-tenancy NoSQL store with workload-aware resource reservation. Parallel Comput 58:76–89. https://doi.org/10.1016/j.parco.2016.06.003

Zhang C, Lu Y (2021) Study on artificial intelligence: the state of the art and future prospects. J Ind Inf Integr 23: 100224, https://doi.org/10.1016/j.jii.2021.100224

Zhang Q, Wang T, Huang K, Chen F (2021) Efficient dispatching system of railway vehicles based on internet of things technology. Pattern Recogn Lett 143:14–18. https://doi.org/10.1016/j.patrec.2020.12.011

Chapter 9
Shaping the Safety Culture in Rail Transport by Creating and Implementing Digital Solutions

Adam Jabłoński

Abstract *Purpose*: The aim of the chapter is to present key assumptions for shaping the safety culture in rail transport by creating and implementing digital solutions.

Design/Methodology/Approach: The methodology for this chapter is a critical literature review, with theoretical and practical reflections aiming to identify aspects that are responsible for shaping a new approach to the safety culture of digitalized railways.

Findings: This chapter identifies the place and role of the safety culture of railway enterprises which undergo digital transformation.

Research limitations/Implications: The limitations of the analytical and evaluation study result from the difficulties in determining issues which are most important when it comes to fully explaining the determinants of shaping safety culture, set against the background of the determinants of the digitalization of the processes of rail transport functioning.

Practical implications: The practical nature of the content described was focused on critical analysis and examples and on indicating the assumptions for the model of shaping safety culture through the digital transformation of the functioning of railway enterprises in terms of person-to-person, person-to-machine, and machine-to-machine relationships.

Originality/Value: The originality of the chapter lies in the model of shaping safety culture vis-à-vis the interfaces of railway system operators and the growing role of automation and robotization of processes. The original value also comes from the justification of the premise that safety is a social factor and is crucial for the creation of the sustainable business models of railway enterprises.

9.1 Introduction

Safety culture is a subculture of the entire organizational culture of railway enterprises, which, as high reliability organizations, define organizational and human behavior mechanisms by shaping a positive climate and a positive safety culture. This

A. Jabłoński and M. Jabłoński, *Digital Safety in Railway Transport—Aspects of Management and Technology*, Springer Series in Reliability Engineering,
https://doi.org/10.1007/978-3-030-96133-6_9

culture becomes a platform for railway enterprises to achieve their economic objectives in a safe manner fulfilling technological and social requirements. It stimulates management intentions, the logic of structured principles of management systems adopted for the implementation of projects, and continuous main, essential, management and supporting processes. Finally, safety culture is becoming a platform for dialogue and communication between a wide range of the stakeholders of railway enterprises. It is becoming a source of building the intellectual capacity of railway enterprises in terms of safety and reliability; the absence of disasters, major accidents and railway accidents should be the result of this potential. In this dimension, safety culture is a specific competence bridge between the clashing objectives, needs and expectations of railway enterprises. A well-shaped safety culture enables the creation of the ecosystem of safe railways based on the same values, principles and methods.

9.2 Railway Enterprises as High Reliability Organizations

High reliability organizations (HROs) refer to organizations that usually operate in hazardous environments where the consequences of process failures are extremely high (Baker et al. 2006). These organizations operate in dangerous conditions, but have less than their fair share of adverse events, take care of safety at the highest level and take a special approach to its implementation (Sutcliffe 2011). High reliability organizations have a distinctive attribute of capability and they operate almost flawlessly in dangerous, complex and uncertain environments (Enya et al. 2019). A digital HRO is an organization that builds services and solutions to protect IT systems from malware or malware threats. High reliability is crucial due to the serious threats of malware to fault-free digital technology (Salovaara et al. 2019). When conducting a broad analysis of the status of railway enterprises in this context from the perspective of management and quality sciences, it can be assumed that they are so-called high reliability hybrid organizations. The concept of hybridity implies a common and concurrent achievement of economic and social objectives (passenger and freight rail transport has a crucial social and economic impact). Reliability is related to the fact that most railway enterprises are reliable, but if a railway disaster or a serious railway accident occurs, both social and economic costs (including loss of human life and health) are disproportionately high. Such organizations maintain full situational awareness, knowing that unexpected events may begin during routine operations arising from a hidden system state or interaction that are usually dormant but quickly turn into a bigger problem if they are not recognized and prevented at an early stage of their development (Cantu et al. 2021). High reliability organizations are an important approach to preventing and mitigating crises. Resilience is a term that has been used to describe the ability of these organizations to react or "rebound" from unexpected, surprising or destructive events (Grabowski and Roberts 2019). Error absorption is a feature of a high reliability organization that allows for an active search for signals in their systems that could indicate this error potential (Agwu et al. 2019). Therefore, real-time problem solving becomes crucial (Le Coze

2019b). Faced with the challenges of controlling unforeseen threats, high reliability organizations should always make every effort to maintain their resilience, including by using safety culture through the involvement of organizational resources which promote a proactive safety system to prevent the recurrence of adverse events (Hollnagel et al. 2006). From this perspective, a factor is emerging that strengthens the implementation of the principles and objectives of the high reliability organization, i.e. the safety climate and culture created, built and developed therein.

9.3 Organizational Culture of Railway Enterprises and Digital Transformation Processes

The concept of safety culture is characterized by high complexity (van Nunen et al. 2018). Safety culture is now an essential concept in building a positive climate in an organization that is conducive to improving safety. For many years this concept has been developed in high reliability organizations in, for example, the nuclear, rail, oil and gas sectors (Le Coze 2019a). When considering the mechanisms of modern management of railway enterprises from the perspective of organizational behavior, it is crucial for them to shape the organizational culture and the safety subculture, which have already been developed by means of the assumptions of digital transformation. This applies, inter alia, to the principles of communication, building human and inter-organizational relationships and shaping the intellectual capital of the organization towards creating safety assumptions and requirements, as a fundamental attribute of the high reliability organization. It is therefore vital to identify factors which influence the creation of safety culture in order to understand the overall safety culture and the promotion of proactive changes in the business safety culture (Wang and Liu 2012). An intriguing question can be asked that is aimed at developing safety culture in high reliability organizations: "Why do our experienced employees (sometimes) work dangerously?" (Farrington-Darby et al. 2005). Hence, it should be remembered that the railway sector is a regulated system whereby the relationship between safety culture and safety performance is well documented in different industries and countries, and regulators in different industries are increasingly integrating safety culture into the assessment of the safety level of specific companies (Nævestad et al. 2019). Therefore, in terms of railway traffic safety, it is important to determine the maturity of the organization. The available evidence clearly implies that enterprise maturity models are a significant means by which to assess safety culture in organizations (Pinto Goncalves Filho and Waterson 2018).

The maturity of enterprises emerges from the concept of sustainable business models, which are a challenge in the context of the dynamics of digital transformation processes. Something mature in the analog formula may not be mature in the digital economy. These organizational and social phenomena pose new challenges for managers of today's businesses. Changing expectations and evolution on the web and the digital economy imply an important need for a broad and open discourse.

This important discussion is necessary to develop new perspectives based on rigorous scientific and theoretical foundations to identify the conditions for business management and business models in the context of sustainable development. The sustainability context includes the balance of numerous measures (Jabłoński et al. 2020). The maturity of organizations supported by the assumptions of sustainable development in this sustainable rail transport supply chain in the context of the digital economy should foster a positive safety culture. This also applies to investment processes, especially in the implementation of systems based entirely on digital technologies (Jabłoński and Jabłoński 2021a, 2021b). The social impact of these investments is vitally important in terms of railway traffic safety. The need to take into account the social impact on strategies to improve the safety culture of railway enterprises emerges.

9.4 Factors Determining the Safety Culture of Railway Enterprises Towards Digital Solutions

A strong interest in how information technologies (IT) change the nature of work in organizations can be observed, which leads to the reconfiguration of accepted working practices (Morton et al. 2020). From this perspective, IT solutions are being sought to support and reduce errors that can be caused by the human factor in the rail transport sector. This may include the Real Time Decision Support System (RT DSS) for railway nodes, which should be able to assist the dispatcher in making reasonable decisions while solving operational problems. Nowadays, these solutions often fully depend on the human factor, which is insufficient for the efficiency of railway node management (Galadíkováa and Adamko 2021). In this context, factors which determine the development of safety culture in railway enterprises towards digital solutions are emerging. Undoubtedly, the first factor which strongly creates the above elements is the leadership of managers, including digital leadership. These open, modern and creative managers of railway enterprises are the ambassadors of dynamically implemented digital transformation. Another especially important factor is the so-called team integration, as it is people who implement innovation and digital solutions. Exploiting the effects of synergy, symbiosis, symmetry, integration, adaptation and consolidation of the already existing IT infrastructure, the interpenetration of digital and analog processes and project work in teams of people with different competencies and from different company departments are the foundations of the effective operation of railway enterprises. Another factor is the value added from the data obtained in the digital system. This has a tangible benefit for building safety culture reinforced by information about the facts and forecasts. These factors strongly strengthen the safety climate, and the knowledge gained through the digital ecosystem supports dynamic decision-making processes, focusing the digital strategy on the fulfillment of safety criteria. It is important to pay attention to the mechanisms of shaping safety culture from the perspective of digital economy determinants. In this cognitive perspective, it is important to define the place and role of digital transformation in shaping the safety culture of high reliability organizations (Jabłoński and Jabłoński 2021a, 2021b). Digitalization creates new opportunities to shape the business ecosystem in the rail transport sector (Jabłoński and Jabłoński 2020). The digital transformation dynamically introduced both in the railway system itself—i.e. the railway network and the configuration of the business models of railway enterprises—contributes to the development of safety culture and poses a significant challenge for railway enterprises. Safety culture based on the digital world determines crucial changes in person-to-person, person-to-machine and machine-to-machine interfaces. These changes are a challenge for the managers of railway enterprises. The relevance of social factors is therefore emerging as a key driver for the digitalization of the business models of railway enterprises. The social approach to shaping the nature of business is an activity which is particularly visible in the rail transport sector. This is affected in particular by the criteria required to clarify the place and role of relationships between railway traffic participants and supporting processes in the context of railway traffic safety. This determines interdependencies in the structure of a complex value added chain based on cooperation between many entities and individuals (Jabłoński and Jabłoński 2019). The shaping of safety culture in railway enterprises will therefore evolve towards digital solutions along with the progress currently being made in this sector. Digitalization will significantly change the understanding of human behavior towards the robotization and automation of rail transport processes.

9.5 Shaping Safety Culture Through the Digital Transformation of Railway Enterprises—A Model

The issue of shaping safety culture in rail transport is not straightforward. This is primarily due to the complexity of the processes conducted and the uncertainty that is a constant attribute of the railway ecosystem. It is therefore interesting to answer the key question of how safety culture can be created through the digital transformation of railway enterprises. In the context of digital transformation, the operational conditions of the railway system are changing in terms of operators' functional interfaces with technical equipment and systems. The multitude of digital solutions means that most processes happen automatically. For example, local control centers, which provide support for a large number of sections of railway traffic, allow railway traffic to be controlled in remote locations by means of control panels and video solutions. This makes most processes almost entirely dependent on the reliability of railway traffic control devices and the protection of collision routes. In this situation, human beings are less likely to generate human error compared to traditional railway solutions, albeit human error cannot be completely eliminated. Safety culture as a set of rules and behaviors of people within a socio-technical system based on honesty, responsibility and professionalism requires reconfiguration in view of the challenges of the digital economy. The role of people in the technical system needs to be redefined. In digital systems, automation supports the decision-making processes of the operator, indicates the boundaries of its tasks and offers full support in terms of the information and decision-making processes. Safety culture must be linked to defining the limits of the organizational responsibility of railway employees, taking into account the scope of operation of digital economy systems, as in many situations they have established space for human participation - one who accepts, legitimizes or blocks the functioning of these electrical, electronic and programmable systems. Then the principles of concept application and functional safety will be crucial to shaping safety culture.

The specification of the requirements for the E/E/PE safety function should include:

a. a description of all safety functions necessary to achieve the required functional safety, which should, for each safety function:
   - provide exhaustively detailed requirements sufficient to design and develop E/E/PE safety-related systems,
   - include the means by which E/E/PE safety-related systems are designed to achieve or maintain a safe EUC,
   - specify whether and during what periods continuous control is required or not when the EUC is attained or maintained safely, and
   - specify whether the safety function is applicable to an E/E/PE safety-related system working in types of work: on a rare recall, frequent or continuous recall.
b. the response time reached (i.e. the time needed to perform the safety function),
c. interfaces between the E/E/PE safety-related system and the operator that are necessary to achieve the required level of functional safety,
d. all information pertaining to functional safety that may affect the design of the E/E/PE safety-related system,
e. all interfaces between the E/E/PE safety-related system and other systems (inside or outside the EUC) necessary for functional safety (EN 61,508–1:2010).

The definition of the system under consideration then includes the following: key subsystems/equipment, architecture and expected behavior, interfaces and operational environment, safety requirements, the definition of the system configuration/system version which the case of safety relates to, reference to input safety requirements and related risk assessment analyses. This approach unambiguously demonstrates the importance of locating operators who perform tasks critical to railway traffic safety in relation to specific hardware and software configurations to fulfill specific functionalities and achieve railway system performances, taking into account the still crucial role of people in supervising the work of machinery, equipment and more broadly technical systems. Figure 9.1 presents a model of shaping safety culture through the digital transformation of the functioning of railway enterprises in terms of P2P, P2M and M2M relationships.

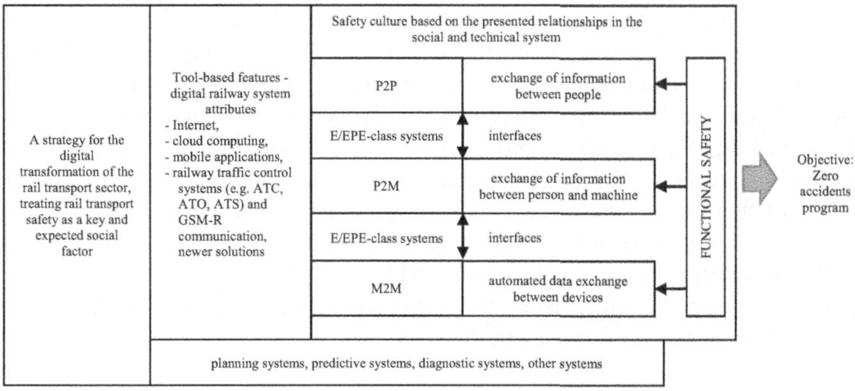

Fig. 9.1 Model of shaping safety culture through the digital transformation of railway enterprises in terms of P2P, P2M and M2M relationships (*Source* Own study)

The presented model is based on the assumption that the digital safety culture is based on relationships between people (P2P), person-to-machine relationships (P2M) and between machines themselves (M2M). All these relationships relate to electrical/electronic and programmable systems. They may, of course, be limited to analog systems. The main task of such an approach is the effective and efficient exchange of data in these relationships. The awareness of the need to improve these relationships for the purpose of shaping a new safety culture should result from the full understanding of the features and attributes of the digital railway system in a tool-based approach, which takes into account basic tools and concepts such as the Internet, clouds, mobile apps, modern railway traffic control systems such as ATC, ATO, ATS and others, as well as digital communication systems based on the GSM standard. The whole is based on the functioning of planning, predictive, diagnostic systems, etc. To achieve this, the strategy for the digital transformation of the rail transport sector should consider the safety of the railway system as a crucial and expected social factor, i.e. every railway stakeholder expects railways to be safe, which affects citizens' sense of safety and increased confidence in railways.

9.6 Conclusions

The person-to-machine interface stems from the assumptions of cybernetics theory. In the age of digitalization, the role of these interfaces is growing significantly, especially when solutions are based on the leading role of the railway system operator. Modern solutions based on full robotization and automation, eliminating human participation, put humans in a completely different role than before. The approach to safety culture from such a perspective must be significantly modified. Digitalization provides the basis for a new paradigm of digital safety management, whereby safety culture will play a key role. This culture results from the level of commitment of the management of railway enterprises in ensuring the safe integration of people's behavior towards technical systems based on digital solutions. Digitalization influences changes in the approach to shaping safety culture and the supervisory and executive role of people in the process of the digital management of rail transport processes. Safety culture forces the need to continuously monitor the correctness of the interaction of human activities with the operation of machines. The system will be safe when it is properly designed, operated and safely changed. This is facilitated by choosing such forms and management styles so that these difficult-to-capture relationships between digital systems and people are identified and used in the process of improving the level of railway traffic safety.

References

Agwu AE, Labib A, Hadleigh-Dunn S (2019) Disaster prevention through a harmonized framework for high reliability organisations. Saf Sci 111:298–312. https://doi.org/10.1016/j.ssci.2018.09.005

Baker D, Day R, Salas E (2006) Teamwork as an essential component of high reliability organizations. Health Serv Res 41(2):1576–1598

Cantu J, Gharehyakheh A, Fritts S, Tolk J (2021) Assessing the HRO: tools and techniques to determine the high-reliability state of an organization. Saf Sci 134:105082. https://doi.org/10.1016/j.ssci.2020.105082

EN 61508–1:2010, Functional safety of electrical/electronic/programmable electronic safety-related systems - Part 1: General requirements

Enya A, Dempsey S, Pillay M (2019) High reliability organisation (HRO) principles of collective mindfulness: an opportunity to improve construction safety management. In: Arezes P (eds) Advances in safety management and human factors. AHFE 2018. Advances in intelligent systems and computing, 791. Springer, Cham. https://doi.org/10.1007/978-3-319-94589-7_1

Farrington-Darby T, Pickup L, Wilson JR (2005) Safety culture in railway maintenance. Saf Sci 43:39–60. https://doi.org/10.1016/j.ssci.2004.09.003

Galadíkováa A, Adamko N (2021) Simulation-based methods to support the real-time management of railway nodes. Trans Res Proc 55:1345–1352. https://doi.org/10.1016/j.trpro.2021.07.119

Grabowski M, Roberts KH (2019) Reliability seeking virtual organizations: challenges for high reliability organizations and resilience engineering. Saf Sci 117:512–522. https://doi.org/10.1016/j.ssci.2016.02.016

Hollnagel E, Woods DD, Leveson NC (eds) (2006) Resilience engineering: concepts and precepts. Ashgate Publishing Limited, Aldershot, UK

Jabłoński A, Jabłoński M (2020) Social perspectives in digital business models of railway enterprises. Energies 13(23):6445. https://doi.org/10.3390/en13236445

Jabłoński A, Jabłoński M (2021) Impact investing in digital business models. Energies 14(18):5785. https://doi.org/10.3390/en14185785

Jabłoński M, Jabłoński A (2019) Social factors as a basic driver of the digitalization of the business models of railway companies. Sustainability 11(12):3367. https://doi.org/10.3390/su11123367

Jabłoński M, Jabłoński A (2021) Shaping the safety culture of high reliability organizations through digital transformation. Energies 14(16):4721. https://doi.org/10.3390/en14164721

Jabłoński M, Timmers P, Sarkis J (2020) Sustainability in business models in the network economy. Electron Mark 30:675–678. https://doi.org/10.1007/s12525-020-00444-1

Le Coze JC (2019) How safety culture can make us think. Saf Sci 118:221–229. https://doi.org/10.1016/j.ssci.2019.05.026

Le Coze JC (2019) Vive la diversité! High reliability organisation (HRO) and resilience engineering (RE). Saf Sci 117:469–478. https://doi.org/10.1016/j.ssci.2016.04.006

Morton J, Wilson AD, Cooke L (2020) The digital work of strategists: using open strategy for organizational transformation. J Strat Inf Syst 29:101613. https://doi.org/10.1016/j.jsis.2020.101613

Nævestad T-O, Hesjevoll IS, Ranestad K, Antonsen S (2019) Strategies regulatory authorities can use to influence safety culture in organizations: lessons based on experiences from three sectors. Saf Sci 118:409–423. https://doi.org/10.1016/j.ssci.2019.05.020

Pinto Goncalves Filho A, Waterson P (2018) Maturity models and safety culture: a critical review. Saf Sci 105:192–211. https://doi.org/10.1016/j.ssci.2018.02.017

Salovaara A, Lyytinen K, Penttinen E (2019) High reliability in digital organizing: mindlessness, the frame problem, and digital operations. MIS Q 43(2):555–578. https://doi.org/10.25300/MISQ/2019/14577

Sutcliffe KM (2011) High reliability organizations (HROs). Best Pract Res Clin Anaesthesiol 25:133–144. https://doi.org/10.1016/j.bpa.2011.03.001

van Nunen K, Lic J, Reniers G, Ponnet K (2018) Bibliometric analysis of safety culture research. Saf Sci 108:248–258. https://doi.org/10.1016/j.ssci.2017.08.011

Wang C-H, Liu Y-J (2012) Omnidirectional safety culture analysis and discussion for railway industry. Saf Sci 50:1196–1204. https://doi.org/10.1016/j.ssci.2011.12.031

Chapter 10
Monitoring Safety in Rail Transport by Means of Digital Economy Solutions

Adam Jabłoński and Marek Jabłoński

Abstract *Purpose*: The purpose of this chapter is to present key objectives of monitoring rail transport safety by means of digital economy solutions.

Design/Methodology/Approach: The methodology for this chapter is a critical literature review, with theoretical and practical reflections aiming to identify aspects that are responsible for the comprehensive monitoring of rail transport safety by means of digital economy solutions.

Findings: This chapter identifies digital solutions in the railway safety monitoring process as well as the role of data transmission systems in ensuring the expected results in terms of digital safety.

Research limitations/implications: The limitations of the analytical and evaluation study result from the difficulties in identifying issues which are of the utmost importance to fully explaining the determinants of railway safety monitoring in view of the progress made in the implementation of innovative digital solutions aimed at improving railway safety.

Practical implications: The practical nature of the content described focused on conducting critical analysis and examples and identifying assumptions for methods and techniques for monitoring rail transport safety in the context of digitalization processes.

Originality/Value: The originality of this chapter lies in the proposed model of the digital monitoring of rail transport safety.

10.1 Introduction

Safety and its principles are based on the continuity of operation, which is determined by risk attributes; the controlled risk is a factor which ensures system safety. In the case of the rail transport system, the continuity of action requires permanent measurement and monitoring. This measurement provides a source of information about the actual conditions of existence of the technical system at the time of real decision making in said system. It should be noted that from this perspective, the

A. Jabłoński and M. Jabłoński, *Digital Safety in Railway Transport—Aspects of Management and Technology*, Springer Series in Reliability Engineering,
https://doi.org/10.1007/978-3-030-96133-6_10

digital economy determines the new and unique measuring and monitoring mechanisms in the rail transport system. The feedback from measurement and monitoring is much faster and more reliable, clearer and easier to interpret. Thus, in this context, digital solutions provide a new shape of a digital measuring system with multiple cognitive perspectives that open up a space for feedback on processes, projects and transport systems. In addition, monitoring and measurement are a tool for preventing, correcting and improving rail transport safety management systems, and output data creates crisis prevention rules.

10.2 Theoretical and Practical Assumptions for Monitoring Rail Transport Safety

Appropriate measuring, monitoring, control, controlling, auditing, and reporting mechanisms can be used to supervise processes that are critical to rail transport safety. The measurement is individual; the monitoring is continuous and is the sum of the measurements taken in a given time function. Control is post-factum, i.e. mainly focused on the past, while controlling is post-ante and concerns the assessment of future-oriented mechanisms. The audit, on the other hand, is a photograph of reality, an active search for compliance in the evaluated processes, and a systematic, independent process related to the verification of whether the set requirements are fulfilled. Reporting is an aggregation of previously assessed or verified data or findings. All these assessment and verification tools which are applicable to railway enterprises should be supported by digital solutions. Such a constructive comparison of analog solutions with digital solutions can give a multidimensional value in terms of providing reliable data, information and knowledge about the safety status of a railway entity. It is worth noting that monitoring requires an evaluation of the entire railway system and is related to the development of a complex measurement system consisting of an aggregated set of indicators that create coherent strategic information for rail transport safety creators. It should be noted that monitoring takes place under supervised conditions when processes are controlled and countermeasures occur in the event of disruptions to the railway system related to disasters, serious accidents, accidents or incidents.

10.3 Methods and Techniques for Monitoring Rail Transport Safety

Rail transport safety monitoring is based on the use of diverse methods and tools used to maintain the level of safety at the expected level. Any deviation from the established plan is treated as a breach of the established safety requirements. According to the Commission Regulation (EU) No 1078/2012, monitoring means the arrangements put in place by railway undertakings, infrastructure managers or entities in charge of maintenance to check their management system is correctly applied and effective. In accordance with that Regulation, it is necessary to check and thereby ensure:

- the correct application and the effectiveness of all the processes and procedures in the management system, including the technical, operational and organizational risk control measures. In the case of railway undertakings and infrastructure managers, checking will include technical, operational and organizational elements that are necessary for the issue of the certification or authorization.
- the correct application of the management system as a whole, and whether the management system achieves the expected outcome;
- the identification and implementation of appropriate preventive, corrective, or both types of measures, if any relevant instance of non-compliance with safety requirements is detected (Commission Regulation (EU) No 1078/2012).

The safety monitoring model legally required by railway undertakings, infrastructure managers and entities in charge of maintenance is presented in Fig. 10.1.

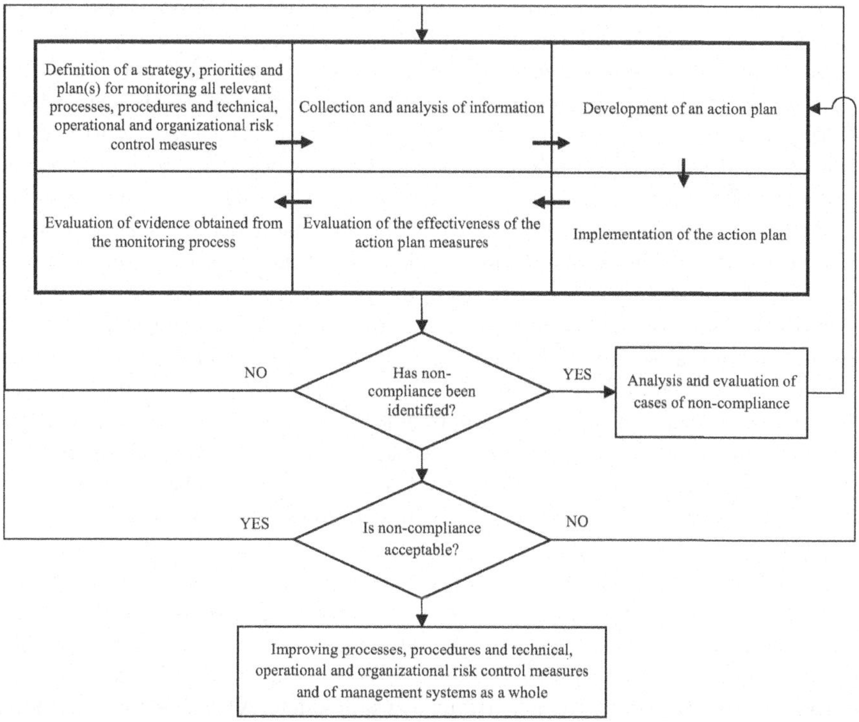

Fig. 10.1 Rail transport safety monitoring model (*Source* Own study based on Commission Regulation [EU] No 1078/2012 since 16 November 2012 on a common safety method for monitoring to be applied by railway undertakings, infrastructure managers after receiving a safety certificate or safety authorization and by entities in charge of maintenance)

Monitoring is a crucial aspect of rail transport safety management as it detects any deviations, limitations and errors that may result in adverse incidents during railway system operation.

The safety management system, which initiates and formalizes the safety monitoring process, is then used to provide all holistic information on the functioning of the railway system and on the activities of railway enterprises which perform different functions.

10.4 Digital Solutions in the Rail Transport Safety Monitoring Process

Monitoring safety by means of digital solutions is part of the overall digital transformation of railways. It is related to the use of Big Data, which, in the context of the specificity of the functioning of the rail transport sector, is easy to obtain. Data on the operation of individual components of railway infrastructure and transport work is in a large number of locations and is recorded in different ways. These include the registers of defects, failures and accidents, consignment note data, train configuration data—train compositions, data on set block sections of the route, data from the video surveillance system e.g. observation of end-of-train signals, data related to the use of different detectors detecting the passage of trains (e.g. RFID gates), data on the identification of rolling stock emergency states, data on monitoring the time and mileage of railway vehicles, etc. A holistic approach to railway traffic safety monitoring and valuation should be based on making full use of digital economy solutions in all possible and economically viable areas of railway enterprise operations in the context of the railway system itself and the operational processes conducted by railway market operators as part of their business models. Therefore, a strategy for the integration of information resources should be pursued to create data warehouses which, depending on the database management model to be implemented (one shared database or distributed heterogeneous data repositories), will be effective and efficient to ensure continuous access to data and information for decision making in the area of railway traffic safety. Figure 10.2 presents the assumptions for the implementation of the strategy for the integration of railway traffic safety data resources.

The presented model identifies two basic approaches to managing railway traffic safety data resources. The first is based on the development and implementation of a single shared database related to railway traffic safety. This is a target strategy. The second solution is the identification and integrated management of heterogeneously distributed data repositories related to railway traffic safety. As regards the integration of distributed data into the central database, this integration will be horizontal. If there is no central, horizontal railway traffic safety database, the integration is vertical on the "each with each" basis. In this way, Big Data can be created and used to monitor railway traffic safety. The architecture of a database system that collects

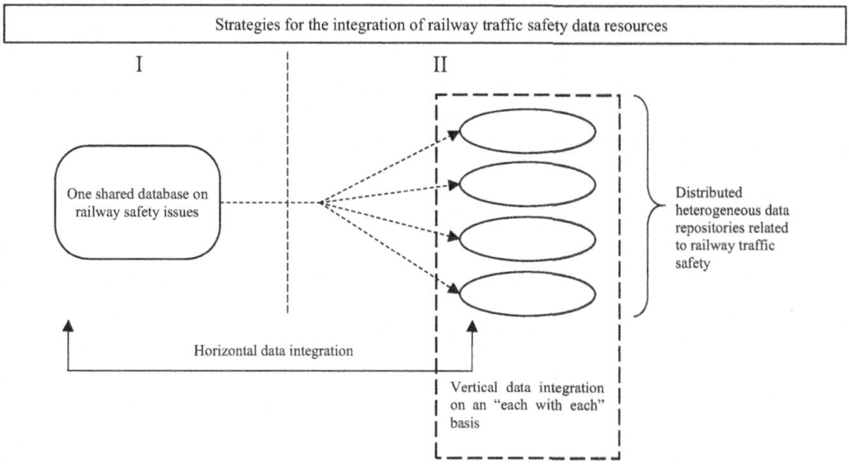

Fig. 10.2 Strategies for the integration of railway traffic safety data resources (*Source* Own study based on Put, D. [2012]. Integracja heterogenicznych zasobów informacyjnych w organizacyjnych dynamicznych [Integration of heterogeneous information resources in dynamic organizations], Kraków: Wydawnictwo Uniwersytetu Ekonomicznego w Krakowie, p. 67)

data from multiple sources allows for the identification of weaknesses and quick decision making.

10.5 Digital Monitoring Model for Rail Transport Safety

Railway traffic safety monitoring processes within the framework of the current rapid digital transformation should develop towards complex database systems with an integrated structure. Figure 10.3 presents the structure of the data management system and the system for monitoring tasks and outcomes in the area of railway system safety. These systems should collect static data from different datasets, such as the European Agency for Railway datasets, e.g. safety alerts, or National Safety Authorities databases. They may also come from standards organizations and institutions such as UIC or VPI. Static data should also be collected on the basis of data from entities, stakeholders and other railway undertakings, infrastructure managers, and entities in charge of maintenance, who have data that may influence the assessment and improvement of the safety of railway enterprises, such as data on faults and failures of users of the same equipment, railway infrastructure or railway vehicles. This data is also of great cognitive significance and is static.

Dynamic data is particularly important and significantly helps to improve the quality of railway traffic safety monitoring processes. Dynamic data will be derived from the effects of the work of sensors built into the various areas of railway system operation. Dynamic data will also result from intelligent solutions built into processes

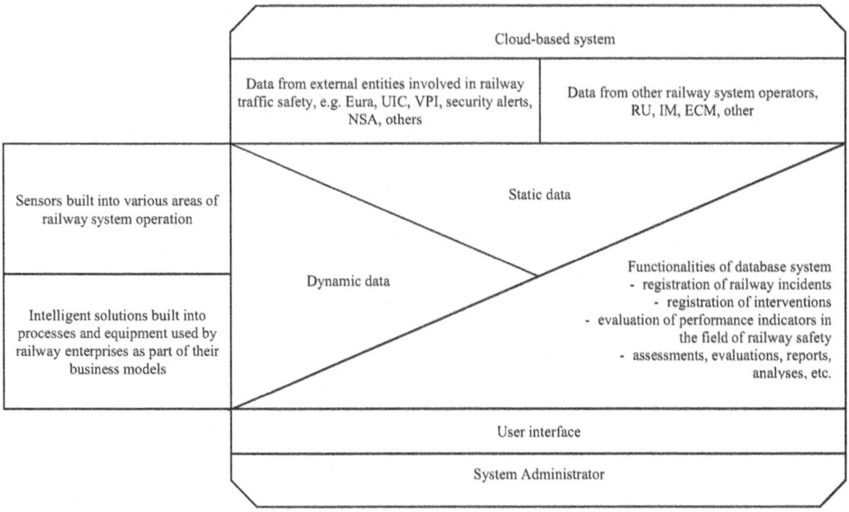

Fig. 10.3 Structure of the data management system and the system for monitoring tasks and outcomes in the area of railway system safety (*Source* Own study based on Sangiorgio, V., Mangini, A.M., & Precchiazzi, I. [2020]. A new index to evaluate the safety performance level of railway transport systems. Safety Science 131, 104,921, doi:10.1016/j.ssci.2020.104921, p. 3)

and equipment used by railway enterprises as part of the functionality of their business models. The overall structure of the data management and outcome monitoring system in the area of railway system safety should be equipped with a number of functionalities which ensure data recording and cataloguing and compilation of data sets and should ensure their effective use for decision making pertaining to railway safety. They should have a built-in user interface and an established data administrator. In this way, tasks, processes and events can be monitored that can support a safety management system based on monitoring static and dynamic data which concurrently shapes the digital monitoring model for rail transport safety.

10.6 Data Transmission Systems as an Aspect Subject to Safety Monitoring

The data transmission system is generally a process focused on effectiveness and efficiency with an appropriate configuration of devices consisting of transmitting any data group between the sender (transmitter) and the recipient (receiver) recorded with a specific code that is understandable for both sides and along a specified route, using the appropriate device system. Modern control systems used in rail transport use data transmission between trackside and on-board equipment. On the one hand, trackside safety control devices retrieve data from light signaling devices or directly interlocking lever frames, and on the other hand, on-board safe control devices transmit instructions to the braking system and warn train drivers. The relevant data, which is critical in terms of safety, is transmitted by means of a variety of transmission media including data channels in radio-communication systems. The complete chain, including data generation, data transmission and data processing, must ensure a sufficiently high level of safety integrity - SIL level 4 (Pawlik 2017). Data transmission must be efficient and safe. This is an essential condition for ensuring the safety of data and the flow of information for the purpose of crucial operational decisions. The safety of digital solutions will therefore depend on data transmission safety, as this may lead to technical system errors, threats of third party hacking or distortion of the signal. In accordance with the EN 50,128 standard, measures should be included in the software to prevent or detect errors occurring during the storage, transmission or duplication of executable code or data. Code-identifiable parameters should be encoded. If the safety-related electronic system involves the transmission of information between different locations, the transmission system is an integral part of the safety-related system and it should be demonstrated that end-to-end communication is safe according to EN 50,129. Safety requirements depend on the characteristics of the transmission system. A transmission system is a service used by an application to transmit telegram streams between multiple participants, each of which can be an information source or outlet. Transmission systems are classified in three categories:

- Category 1 covers systems that remain under the control of the designer and are unchanged throughout their lifetime;
- Category 2 covers systems that remain partially unknown or not unchanged, but for which unauthorized access can be excluded;
- Category 3 covers systems that are outside the control of the designer and for which unauthorized access should be included.

Data transmission systems in the rail transport sector should therefore be encoded. Telegram encoding is the transformation of bits within a telegram by using a cryptographic technique in a manner consistent with a key-controlled algorithm, in order to make it difficult to read the data. Encoding does not protect against data corruption (EN 50,129). The structure of the railway system implemented using safety-related and non-safety-related computer applications is presented by Fig. 10.4.

Both safety-related and non-safety related computer applications are used in the rail transport system. Together they form one technical ecosystem. Safety-related and non-safety-related telegrams and information alike are transmitted between applications. In any case, safety-related cryptographic techniques should be used in order to achieve the safety of data and information and in particular the safety of operations. From a practical point of view, different hazards may occur in a fully digitalized railway system, for which risks should be estimated. For example, if the telegram to the train is damaged due to random errors, it may allow the train to enter the occupied section and collide with another train. In addition, intentional attempts may be made to enter the wrong telegram into the system (e.g. by a hacker). A possible means of protecting the telegram against damage is the use of a safety code attached to the user's information in the telegram. By introducing this means of protection, an acceptable

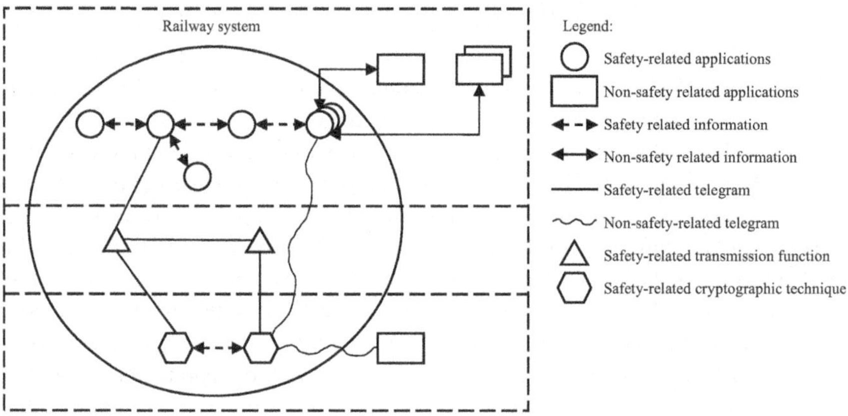

Fig. 10.4 Structure of the railway system implemented by means of safety-related computer applications and non-safety-related computer applications (*Source* Own study based on EN 50,129:2019, Railway applications—Communication, signaling and processing systems—Safety-related electronic systems for signaling)

level of safety can be achieved. Figure 10.5 shows the scope of the assessment of the safety of telecommunications networks which use data transmission.

The identification of the scope of the safety of telecommunications networks covers a wide range of issues. It requires the implementation of appropriate safeguards, diagnostics and data and information protection tools. It also requires the broad identification of risks in the digital rail radio communication system. Such activities contribute significantly to improving the safety of systems based on data transmission.

Digital safety management for data transmission systems therefore requires the implementation of a number of solutions which prevent the occurrence of hazardous events, which should be eliminated by mitigating the risk of their occurrence.

10.7 Conclusions

Safety monitoring requires the use of both analog and digital solutions. In the case of equipment, railway vehicles and technical systems where digital solutions have been used to a large extent, monitoring can be direct with the use of a large number of sensors and performance evaluation dashboards. Such measurement of complex technical systems enables online evaluation to immediately detect weak signals of deviation that require immediate intervention. Such a direct model of monitoring the proper functioning of the railway system in relation to the direct technical structure helps to prevent failures, and any such failures are subject to an immediate response. This eliminates major accidents and disasters. The system of monitoring and visualizing results allows for the better identification and analysis of the causes of adverse

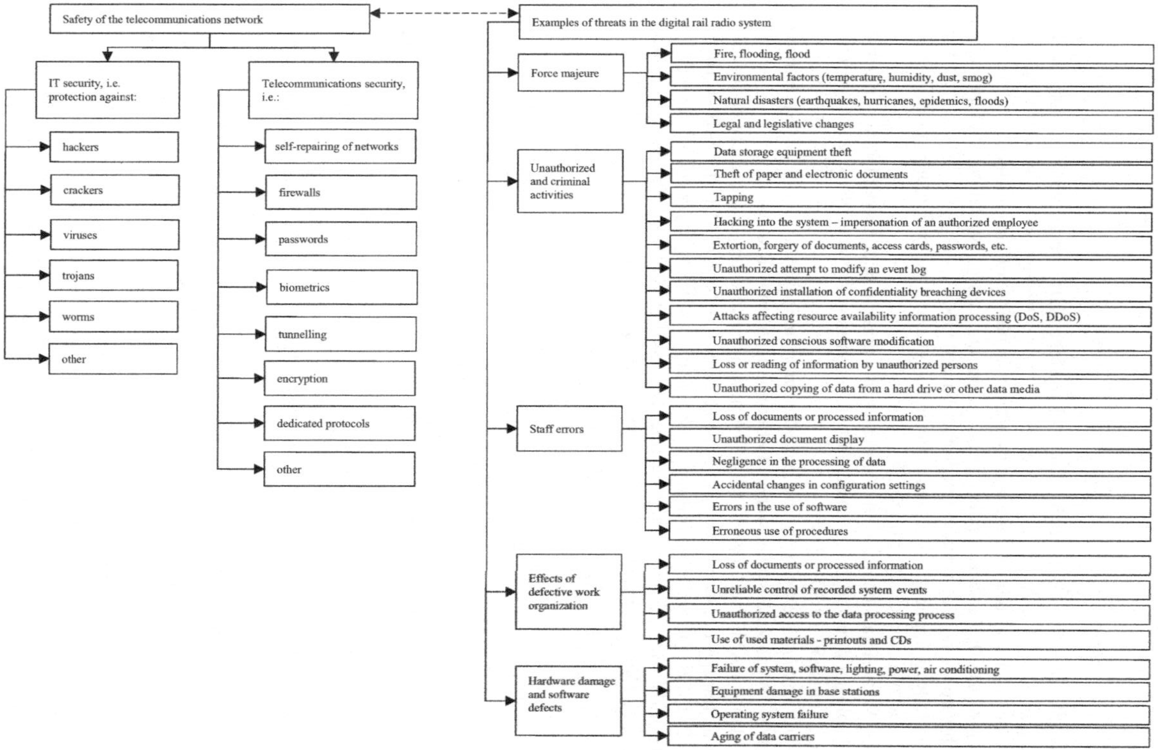

Fig. 10.5 Scope of the assessment of the safety of telecommunications networks using data transmission (*Source* Own study based on Sergejczyk, M. [2018]. Bezpieczeństwo transmisji informacji w sieci cyfrowej radiołączności kolejowej [Safety of the transmission of information over the digital network of radio communications]. Warszawa: Oficyna Wydawnicza Politechniki Warszawskiej, pp. 10, 65–66)

events. Only this approach to measuring and monitoring railway safety makes it possible to significantly reduce risk and improve railway operation processes.

References

Commission Regulation (EU) No 1078/2012 of 16 November 2012 on a common safety method for monitoring to be applied by railway undertakings, infrastructure managers after receiving a safety certificate or safety authorization and by entities in charge of maintenance.

EN 50128:2011/A1:2020, Railway applications - Communication, signalling and processing systems - Software for railway control and protection systems.

EN 50129:2019, Railway applications - Communication, signalling and processing systems - Safety related electronic systems for signalling.

Pawlik M (2017) Wyzwania techniczne oraz ograniczenia prawne podczas odbierania i przekazywania do eksploatacji urządzeń bezpieczeństwa aktywnego. Prace Naukowe Politechniki Warszawskiej, Transport 19:301–318

Put, D (2012) *Integracja heterogenicznych zasobów informacyjnych w organizacyjnych dynamicznych*, Kraków: Wydawnictwo Uniwersytetu Ekonomicznego w Krakowie.

Sangiorgio V, Mangini AM, Precchiazzi I (2020) A new index to evaluate the safety performance level of railway transportation systems. Saf Sci 131. https://doi.org/10.1016/j.ssci.2020.104921

Sergejczyk M (2018) *Bezpieczeństwo transmisji informacji w sieci cyfrowej radiołączności kolejowej [Safety of the transmission of information over the digital network of radio communications]*, Warszawa: Oficyna Wydawnicza Politechniki Warszawskiej.

Chapter 11
Conceptualization and Operationalization of the Rail Transport Digital Safety Model

Adam Jabłoński and Marek Jabłoński

Abstract *Purpose*: The purpose of this chapter is to present the key assumptions for the conceptualization and operationalization of the rail transport digital safety model.

Design/Methodology/Approach: The methodology for this chapter is the process of setting assumptions for the original rail transport digital safety model.

Findings: This chapter identifies safety management methods and tools to support the development of a rail transport digital safety model.

Research limitations/implications: The limitations of the analytical and evaluation study result from the difficulty in identifying issues that are of the utmost importance for the full description of the rail transport digital safety model.

Practical implications: The practical nature of the content described was focused on a scientific discourse aimed at developing the concept of the rail transport digital safety model and its feasibility in the rail transport sector.

Originality/Value: The originality of this chapter lies in the integration of the rail transport digital safety model and the areas of application of digital systems and technologies supporting the digital transformation of rail transport.

11.1 Introduction

The processes of the conceptualization and operationalization of theoretical models are based on the identification of the fundamental variables describing these models and their place and role in the overall configuration of the system of related components. Such an approach in theory and practice allows for the explanation of the relationship between the parameters describing the model and the identification of such components that are most important for the practical impact of a given model on the actual work processes of the system. In terms of rail transport and the issue of railway traffic safety, the creation of model solutions is not easy due to the high level of complexity of the technical, operational and organizational system such as this railway system. A particular difficulty is encountered when dealing with such a broad and interdisciplinary theme of the digitalization of the railway system and the

A. Jabłoński and M. Jabłoński, *Digital Safety in Railway Transport—Aspects of Management and Technology*, Springer Series in Reliability Engineering,
https://doi.org/10.1007/978-3-030-96133-6_11

interdependent digital transformation of the business models of railway enterprises. Rail transport digital safety is a subject that essentially refers to all activities in rail transport in the broadest sense. Hence, the identification of core features, components, tools and systems requires a broad review of literature, verification of past experience from the railway market, but also its own interpretations, conceptual and analytical works, which have been presented in this monograph. The proposals in the chapter are the outcome of the analysis and synthesis of the results of literature research, market observation including trends, case studies of good solutions and specific approaches to the research subject. The chapter proposes the description of the rail transport digital safety model, which aims to build an effective and efficient set of components shaping the configuration of the railway system, the protection of which against faults, failures, incidents, disasters and any other malfunctions are built on the cybersecurity canvas based on technical, legal and organizational and normative solutions.

11.2 Digital Technologies Supporting the Development of a Rail Transport Digital Safety Model

Digital technologies support the development of a rail transport digital safety model. Digital economy technologies generally have different origins. Their development is based on the experience of numerous global economy sectors. Their adaptation to the rail transport sector is as natural as it is to other sectors. The transformation of analog solutions into digital ones is related to the emergence of increasingly advanced business ecosystems based solely on the assumptions and solutions of the digital economy. Due to its technological specificity, rail transport is at the forefront of the search for solutions that, in addition to having innovative features, will be safe and at the same time fulfill the requirements of technical compatibility and safe integration. Therefore, digital technologies undergo numerous tests for a wide range of applications. Their functionality depends on the needs of a given solution as part of the operation of the railway system. The basic scope of the digitalization of rail transport processes includes:

- Computerization of internal processes in companies,
- Automation of administrative services,
- Optimization of transport networks using geographical modelling solutions.

Another element of the digitalization of the sector is the platformization of sales, i.e. the automation of purchasing processes and the use of digital platforms, bringing together different services. The third dimension of the digitalization process is the impact of digital giants on the market. Large global players, which have emerged thanks to technology, are putting pressure on the industry by demanding large-scale transport and logistics services. They also develop their own transport activities within their own capital groups. The final element of digitalization is advanced digital technologies—telematics, solutions based on artificial intelligence, blockchain and process automation. Simultaneously, alternative driving and transport autonomy solutions are being designed and implemented. The requirements and assumptions of Industry 4.0 and 5.0 will allow for even wider use of Big Data technology in transport and logistics systems. Solutions in the telematics of transport systems, which will contribute to the development of artificial intelligence or blockchain-based solutions, are also important for the development of technologies in rail transport. The Internet of Things (IoT), Machine Learning/Predictive Analysis and Artificial Intelligence (AI) are digital technologies that are widely applied to rail transport and the digitalization thereof. Moreover, logistical support for digital technologies, in areas such as tracking and analysis of shipment patterns, anticipating demand and increasing customer satisfaction, better quality and monitoring of maintenance work, allows for the reduction of the number of repairs and downtime, and improving the retention of fleet operators. Predictive maintenance, digital storage and data sharing are developing in particular. These and other digital technologies based on technological advances stemming from scientific and industrial research make the dynamics of changes in the functioning of rail transport systems significant and visible, not only among railway system operators but especially among rail users.

11.3 Safety Management Methods and Tools Supporting Shaping the Rail Transport Digital Safety Model

The use of different safety management methods and tools supports the process of shaping the rail transport digital safety model. Safety management systems implemented by infrastructure managers and railway undertakings are supported by various tools related to so-called hard and soft issues. In the case of digital security, both of these areas apply. The hard methods include risk control measures based on the protection of equipment and systems against failures and third party interference. All forms of data encryption based on technical standards and the best available techniques are the foundation for maintaining the expected level of cybersecurity and process safety, assuming that cybersecurity is obviously an essential part of process safety. Soft methods include the processes of shaping safety culture in relation to risks posed by digital economy solutions, including the education of staff who operate digital devices, staff responsible for planning and ensuring the continuity of operation of these systems, as well as raising awareness of the risks posed by such solutions. Preventive learning should be based on the concept of learning by doing. These two approaches are essential, as together they cover the entire range of impact on the safety of new solutions resulting from the digital transformation of the rail transport sector. The safety and security approach should therefore be considered together, particularly in terms of digital safety. As regards cybersecurity, important issues will include:

- Methodologies for building a cybersecurity program,

- Risk identification, review and estimation,
- Cybersecurity risk management models,
- Systems supporting the risk management process (incident management),
- Risk management in critical infrastructure systems,
- Identification of threats - cybercrime,
- Organizational and process aspects of cybersecurity management, as well as standards in cybersecurity,
- Selected aspects of the administration of information security and the protection of classified information,
- The basics of cryptography and secure information exchange,
- The use of advanced analytical methods in cybersecurity.

In terms of process safety, crucial issues include:

- Methods and techniques of the identification of hazards and the valuation and assessment of operational risk,
- Methods and techniques of change management,
- Methods and techniques of assessing railway incidents,
- Methods and techniques of assessing human errors,
- Methods and techniques of monitoring railway safety,
- Methods and techniques of assessing the correctness of performing tasks critical to railway safety, taking into account processes supported or entirely based on digital solutions,
- Methods and techniques of the implementation of a digital culture strategy for railway traffic safety.

These ranges of the application of digital safety management methods and tools in terms of security and safety provide opportunities for a wide range of the application of these solutions. As regards cybersecurity, tools include an enriched data model from all data sources to strengthen people and artificial intelligence, which provides a safety method such that detections are generated based on a raw Interflow to obtain alerts of highest fidelity threats, dedicated network and security sensors that can be deployed in the form of hardware or software, and the use of cloud computing. In terms of process safety, tools include dynamic fault detection through predictive systems for railway infrastructure and rolling stock maintenance processes, risk management methods such as FMEA, ETA, FTA, HAZOP, and others, methods for assessing human and organizational factors such as the SHELL method to assess the relationship between people and other elements of shaping a safe working environment in the railway system, and others. The combined implementation of methods and tools for digital safety management in terms of security and safety should ensure that the expected level of protection against dangerous events is achieved and that safety management in such a holistic approach is effective and efficient.

11.4 A Rail Transport Digital Safety Model

The digital safety model is a coherent set of devices, technical systems and components supporting a safety management system, which uses digital solutions as a basis for maintaining the expected level of safety. The digital safety model is a combination of digital economy solutions embedded in individual components—the functions of the railway system. It is a set of elements which shapes the configuration of the railway system, the protection of which against defects, failures, incidents, disasters, and any other irregularities are built on a cybersecurity canvas based on technical, legal-organizational and normative solutions. Digital solutions of the digital safety model will be grouped in the following way (Fig. 11.1):

1. Areas of application,
2. Place and role in the railway system,
3. Belonging to a specific family of the technology used,
4. Belonging to a specific function of operation in the safety management system.

As regards the area of application of digital safety, it will cover both the space dedicated to the railway system as a place where railway traffic takes place directly and the indirect areas of impact. The digital economy means that railway systems can be controlled from greater distances by means of different digital applications. These applications shape a digital ecosystem, which also consists of the components of digital business models operated by railway enterprises. The area of the application of rail transport digital safety should be viewed in such a broad way.

Another aspect of this issue is the important role played in management systems by digital safety. Management systems, and in particular safety management systems, force the use of supervision over the risks of rail transport operations. This applies to both technical and operational processes and aspects supporting these processes.

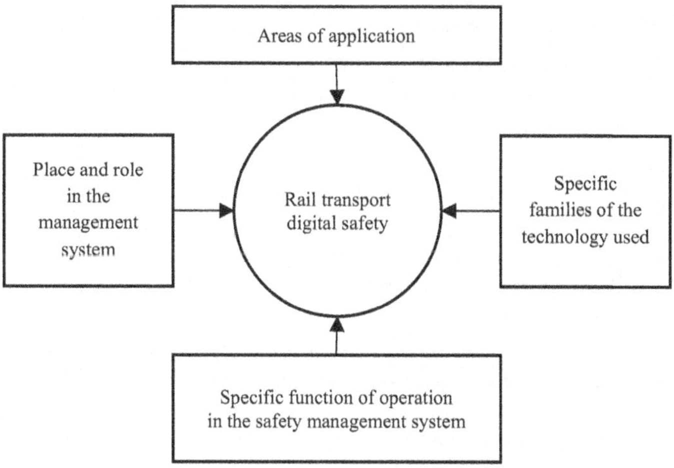

Fig. 11.1 The extent of the impact of rail transport digital safety (*Source* Own study)

Hence, a holistic approach to risk identification also makes digital safety highly dependent on management aspects embedded in the relevant procedures which describe the work processes of the railway system. The role of people in decision making is of the utmost importance. Another element which is important for the effective and efficient management of rail transport digital safety is the identification and understanding of attributes of specific families of technologies used. In recent years, various types of digital technologies have developed dynamically, and are transferred to the rail transport sector. They come not only from families of technologies dedicated to the rail transport sector, such as control systems based on ERTMS assumptions, but also other digital economy solutions such as the sharing economy, the circular economy, Big Data, etc. These "hard technologies" (ETCS, GSM-R) are supported by so-called "soft technologies" (the sharing economy, the circular economy, etc.). In the future, however, they will create a single digital ecosystem together. The final element of the proposed approach is the identification of specific functions of operation in the safety management system. Each application can generate value from its functionality, but it can also generate risks. The safety system should be integrated into the assumptions of the functioning of the railway enterprise safety management system. The identification of hazards within the specific functions of digital applications allows for the development of a coherent hazard register, on the basis of which it is possible to determine an acceptable risk relevant to each functionality of those digital applications. If all functionalities of digital applications are also safe in the area of interfaces with other applications, once all risks have been covered by the safety management system, the railway system can be operated safely.

The key to the identification of digital safety areas is the identification of the types of systems that can operate in a given area, as shown in Fig. 11.2.

For each of the defined areas, the systems where digital safety management methods can be used based on different approaches and solutions have been identified. From a holistic perspective, the railway system will be safe when all areas of action based on the digital economy have assigned protection systems against errors and cyber-attacks. In each of the defined areas of operation of the digitalized railway system, requirements should be defined in three core areas, namely:

1. Safety.
2. Security.
3. Maintenance.

Figure 11.3 presents three core areas of the safety management of the digitalized areas of the railway system functioning.

The configuration of devices, systems and operators by means of digital economy solutions should be safely integrated. In addition to technical safety, safe integration should consider cybersecurity and threats to maintenance processes. The role of the human factor, including the risk of human error, will also be an important issue. The modern approach will also include an innovative environment of railway system operation, taking into account Industry 4.0 and Industry 5.0 concessions based on the full automation of railway systems.

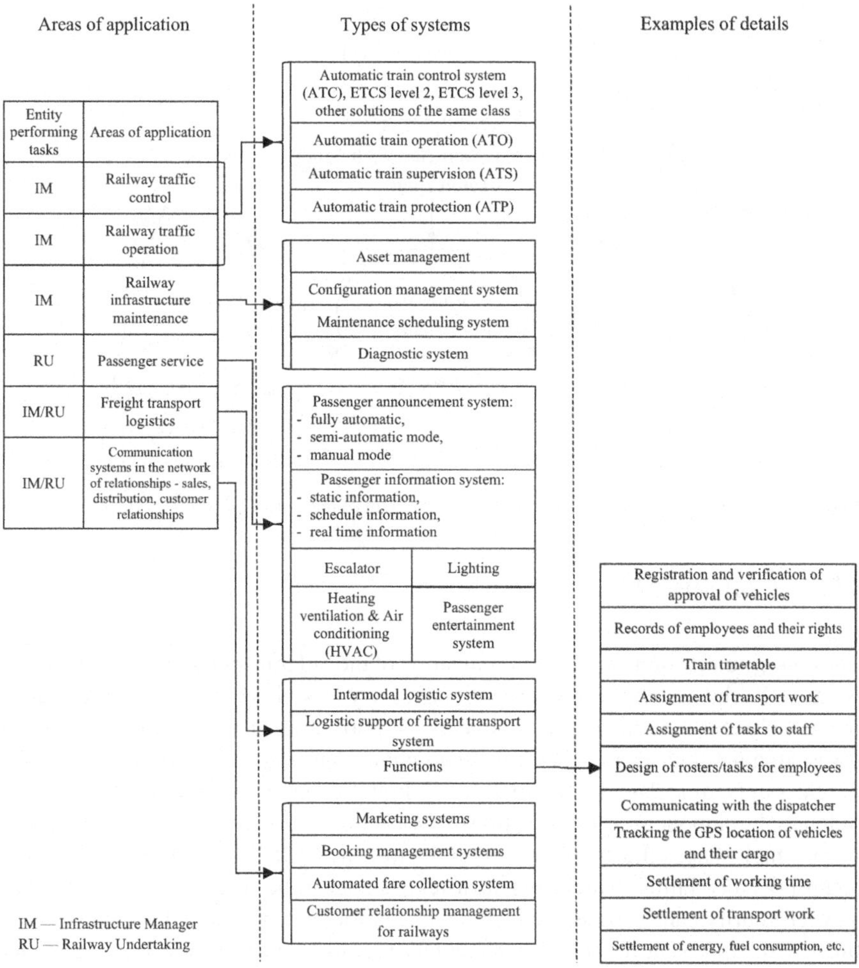

Fig. 11.2 Areas of application of systems supporting the digital transformation of rail transport (*Source* Own study)

Safe integration in such complex systems is essential to ensure that the railway system configuration used fulfills the requirements of technical compatibility and can be operated safely. The responsibility for the safe operation and safe railway traffic management and for the proper control of the associated risks does not lie with a single railway entity. The infrastructure manager and all railway enterprises which operate on its network share this responsibility—each for its part of the system. In a complex environment based on structural and functional interfaces, safe integration should be ensured based on the principle that any system, at any time of its functioning, must be safe even when it is changed—that is to say, throughout the life cycle along with the renewal and modernization phases.

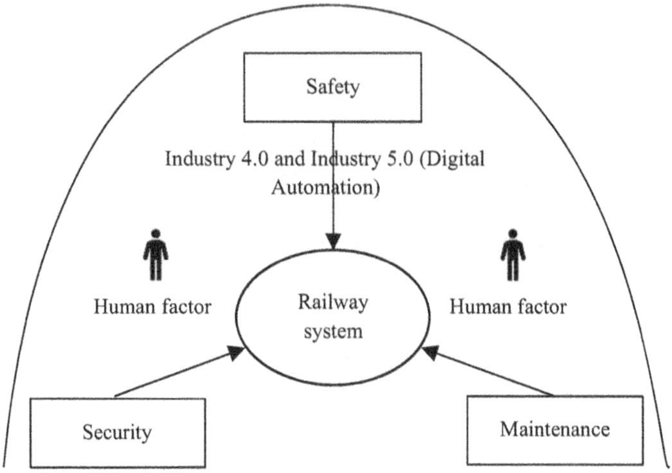

Fig. 11.3 Three core areas of digital safety management of the railway system (*Source* Own study)

11.5 Conclusion

The conceptualization and operationalization of the rail transport digital safety model is based on the assumption that the digital safety model is a coherent set of equipment, technical systems and components supporting a safety management system using digital solutions as a basis for maintaining the expected level of safety. The digital safety model is a combination of digital economy solutions embedded in individual components—the functions of the railway system. It is a set of components which shapes the configuration of the railway system, the protection of which against defects, failures, incidents, disasters, and any other irregularities are built on a cybersecurity canvas based on technical, legal-organizational and normative solutions. This approach to safety management requires consideration in terms of the configuration of the components which shape the railway system together with the supporting digital business models operated by railway enterprises. The blurring and fuzzy boundaries between the business models of railway enterprises and the railway system, often due to their embeddedness in a single digital platform, require the recognition and establishment of new assumptions for railway traffic safety management. The full digitalization of railways means that a new digital safety management paradigm is shaped based on the logic of cybernetics, which the new perspectives of cyber security, technical and process safety and safety management of the railway system treated as a complex socio-technical system derive from.

Chapter 12
The Safety Triad in the Business Models of Railway Enterprises (Management–Digitalization–Socialization)

Adam Jabłoński and Marek Jabłoński

Abstract *Purpose*: The aim of the chapter is to present the assumptions for the proposed safety triad in the business models of railway enterprises in management, digital and social terms.

Design/Methodology/Approach: The methodology for this chapter is the process of setting assumptions for the original safety triad in the business models of railway enterprises.

Findings: This chapter identifies such aspects of rail transport digital safety that shape the safety triad in the business models of railway enterprises, proposed and developed in the course of scientific analyses.

Research limitations/Implications: The limitations of the analytical and evaluation study result from the difficulties in identifying issues which are of the utmost importance for the full identification of all factors responsible for the proposed safety triad in the business models of railway enterprises.

Practical implications: The practical nature of the content described was focused on conducting a scientific discourse aimed at developing the determinants of the concept of a safety triad in the business models of railway enterprises.

Originality/Value: The originality of this chapter lies in the concept of the visualization and interpretation of the interdependent relationship between safety management mechanisms, the digitalization of railway systems and supply chains, and the recognition of the concept of rail transport safety as an important social factor.

12.1 Introduction

Safety in the rail transport system, as indicated repeatedly in the monograph, is interdisciplinary and should be considered in terms of management, technology, and socialization, as well as in the context in the digital economy. The technical sphere of digital safety does not explain the matter in full. This is due to the fact that the key stakeholders of a high level of railway safety are the shareholders of companies, but in particular railway users, who must show the proper respect to the use of this mode of transport. It is also crucial to note that conventional and high-speed rail differ in

A. Jabłoński and M. Jabłoński, *Digital Safety in Railway Transport—Aspects of Management and Technology*, Springer Series in Reliability Engineering,
https://doi.org/10.1007/978-3-030-96133-6_12

their architecture when it comes to their level of digital maturity. This differentiation leads to the division of railway systems into more digitalized and, in many situations. more analog ones. The widespread implementation of safety improvement strategies determines the indication that the weakest link in the railway system is the human factor, while digital solutions, which number among the technical solutions, are characterized by an extremely low failure rate, which significantly affects the level of safety. Looking at railways from the perspective of the digital economy while also taking into account the relationship between the digital business model of railway enterprises, the social role of safety and technical aspects of digitalization seems to be the right way forward. Safety management mechanisms, as well as the digitalization and socialization of the railway traffic safety issue, give rise to a set of related components that have not yet been seen in the literature as a paradigm of a new approach to digital safety management. This approach is proposed by the authors, and is described in this chapter in the form of a safety triad in the business models of railway enterprises. A broad view of the digitalization of rail transport safety should allow for a strategic reflection on this subject by readers of the book, as well as giving a specific tone of understanding of the subject in a somewhat broader context.

12.2 Synthesis of the Simultaneous Implementation of the Analog and Digital Layers of Safety Management Mechanisms in Rail Transport

Rail transport safety management mechanisms in the digital transformation process are significantly modified in some areas. These changes are exceptionally large in some areas, and smaller in others. However, in the context of the implementation of digital solutions, they generally occur throughout the entire area of the functioning of the railway system. The digital transformation, which relates to technical, operational and organizational issues, means that existing mechanisms need to be revised. The issues of the digitalization of processes at different levels of the functioning of railway enterprises relate to the direct and indirect impact on railway traffic safety. All these influences change rail transport safety management mechanisms, as methods, tools and ways of taking action are changing along with the digitalization process. It is important to be aware of the civilization progress and the impossibility of stopping it, which means that rail transport safety managers must constantly learn and look for effective and efficient mechanisms to mitigate risks and improve railway traffic safety performance. Digital technologies and solutions created on the platform which they provide integrate railway systems more strongly than mechanical and relay systems, creating highly complex cyber systems. At present, they not only concern the railway traffic control layer, but cover all processes of the supply chain. It is digitalization that makes it possible to supervise all the processes that make up the supply chain. This concerns the use and diagnostic layer of railway infrastructure, the railway traffic control layer, the railway traffic management layer and the overall layer of railway logistics. Only the understanding of the simultaneous implementation of the analog and digital layers of rail transport safety management mechanisms in the digital transformation of railways can create a new paradigm of rail transport digital safety management. This paradigm should contain a set of new rules and principles of safety management in rail transport as shown on the pages of the monograph. Moreover, rail transport safety management mechanisms should be based on, but also support, the business model of railway enterprises. In many cases, it can already be called a digital business model. This is due to the fact that most of the operations conducted by railway enterprises rely almost entirely on digital solutions. Safety management systems and digital technologies are then based on digital economy safety systems, including in particular electrical/electronic and programmable systems for which functional safety requirements are described in the relevant technical standards. In a modern approach, it is necessary to combine management functions and technical functions into a single integrated whole, which is logical given the conditions of design, construction and operation of modern rail transport systems. The innovativeness of solutions in the rail transport sector also stimulates new needs to build appropriate organizational and technical safeguards and management solutions, which, on the one hand, will enable innovative solutions in rail transport, subjecting them to relevant certification processes, and on the other hand integrating them with already existing solutions, fulfilling the requirements to ensure full technical compatibility and safe integration.

12.3 A Hybrid Rail Transport Safety Management Model Based on the Mechanisms of Analog and Digital Layers

Hybridity is generally something that consists of different, often incompatible elements. A minimum hybrid solution can consist of two different components, solutions, systems, etc., forming one goal-oriented system. When considering a hybrid rail transport safety management model based on analog and digital layers, safety management mechanisms derived from both of these often intertwined areas can be distinguished. It should be noted that the analog layer is a completely non-digital solution or one where the digital solution is marginal. By contrast, a digital solution means a solution that operates entirely in the socio-technical and organizational ecosystem based on digital solutions (Fig. 12.1).

The model shows the location of digital safety management mechanisms in relation to the broader perspective of comprehensive safety management mechanisms in rail transport set against the background of a hybrid perspective. It is assumed that digital safety management mechanisms comprise such railway solutions that are directly built into the railway digital ecosystem, provided that the rules of the full digitalization of operational processes are fulfilled. Thus, digital safety is part of overall rail transport safety. Both areas of safety shape the business models of railway

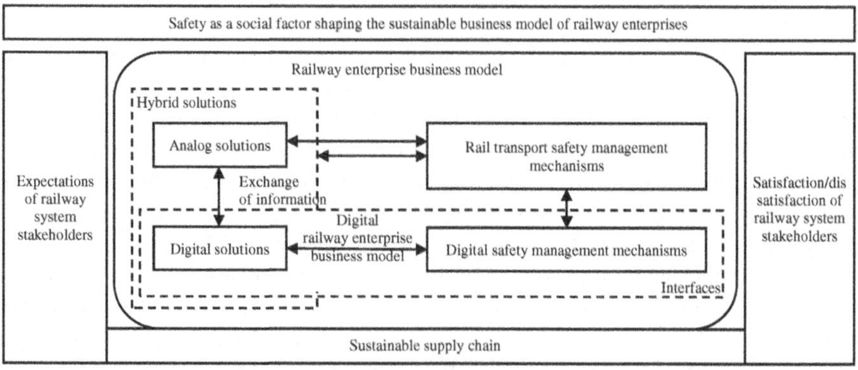

Fig. 12.1 Location of digital safety management mechanisms in relation to the broader perspective of comprehensive rail transport safety management mechanisms set against the background of a hybrid rail transport safety management model (*Source* Own study)

enterprises. The digital business model of a railway enterprise is a company with a very high degree of maturity in the implementation of digital solutions. This will be a level characterized by work in the digital ecosystem. All operational decisions, executive, planning, and diagnostic actions are based on information and decision-making systems supported by digital solutions. It should also be noted that the proposed solutions of a holistic nature, albeit indicating some maturity options in the field of digitalization, are supported by the assumption that safety is an important social value. Stakeholders' needs are focused on the expectation that the railway system will always be safe, assuming that railway enterprises meet the definition of a high reliability organization. This shapes the attributes of sustainable business models of railway enterprises responsible for shaping sustainable supply chains.

12.4 Conceptualization of the Safety Triad in the Business Models of Railway Enterprises

A safety triad embedded in the business models of railway enterprises can be a way of visualizing and interpreting the interdependent relationship between safety management mechanisms, the digitalization of railway systems and supply chains and the recognition of the concept of rail transport safety as an important social factor (Fig. 12.2).

The safety triad presented takes into account the role played by the business model of railway enterprises, which, supported by a positive safety culture, should be organized in such a way that economic activity is conducted in a safe manner, which is an assumption for safety management systems. The triad consists of a set of three interconnected components, namely safety management mechanisms as a set of rules, procedures, principles and methods used to achieve the expected

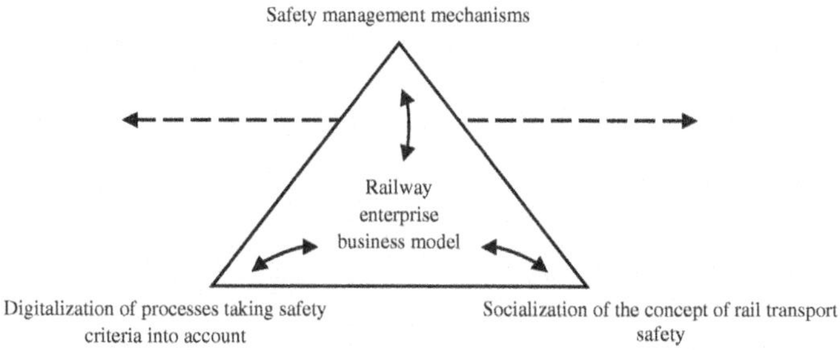

Fig. 12.2 Safety triad in the business models of railway enterprises (*Source* Own study)

level of railway safety. The second component is the digitalization of processes pertaining to rail transport activity, with a particular focus on digital safety. The final component of the triad is the socialization of the concept of railway transport safety, which stems from the expectations of stakeholders in terms of intolerance to railway incidents. The triad responds to the challenges faced by socio-technical systems nowadays in the context of improving digital safety management processes based on the business models of railway enterprises. Supply chains run along the railway system supported by digital economy solutions, the ontological nature of which makes it possible to capture all its links. Planning and prediction in a dynamic system allows for the maintenance of the expected continuity of operation and the filling of all spheres of activity, achieving high effectiveness and efficiency. These parameters are particularly significant in terms of the intermodal level of competitiveness of the rail transport sector in relation to road, maritime, air and river transport.

Chapter 13
Conclusion—Empirical and Methodical Conclusions

Adam Jabłoński and Marek Jabłoński

When summarizing the results of the studies and analyses presented in the monograph, it is necessary to confirm that the objectives of the work have been achieved. The main objective of scientific discussion, which was to present, from multidimensional and holistic perspectives, the rail transport digital safety management mechanisms in management, social and technological terms, has also been achieved. A broad review of the literature, the observation of railway enterprises in this area and the ideas and conclusions developed by the authors confirmed the importance of the issue of rail transport digital safety in social and technological terms. The issue of digital safety is also of particular importance when set against the background of the assumptions of digital business models. This subject is extremely broad, not only in terms of the holistic nature of the safety issue, but in particular the specificity of the functioning of the rail transport sector. In this sector, in the field of two key ontologically separable research spaces, namely safety and security, the decision was made to combine them into a single model called Cyber Operational Safety (COS) the assumptions of which should cover all risks arising from the operation of computer applications directly related to safety and those that support technical, operational and logistical processes in rail transport. In this respect, efforts were made to develop an original approach to clarifying the scientific and practical problem posed. The authors attempted to fill this cognitive gap.

Cognitive goals that included:

1. Analysis of the theoretical assumptions of the mechanisms of creating digital safety in the business models of railway enterprises.
2. Analysis of the theoretical assumptions of identifying the relationship between the digital determinants of the business models of railway enterprises and digital safety management.
3. Analysis of the digital transformation process in rail transport and its impact on safety management processes, treating safety as an important social factor.

A. Jabłoński and M. Jabłoński, *Digital Safety in Railway Transport—Aspects of Management and Technology*, Springer Series in Reliability Engineering,
https://doi.org/10.1007/978-3-030-96133-6_13

4. Analysis of digital innovation management processes in rail transport in terms of improving rail transport safety.
5. Analysis of the impact of digitalization on the development of safety culture in rail transport have been achieved.

The authors' assumption that these issues build the specific nature of the activity of railway enterprises in terms of shaping business models focused on the processes of digital transformation in the business space and the nature of the railway system in which they create, work and operate has been confirmed. In cybersecurity terms, a railway system is a complex system of related components that are interrelated both technologically and socially. The subjective approach is linked to the objective approach. It is the railway market players who shape the railway system, which is oriented towards achieving the primary objective—that is to say, to manage railway traffic in a safe way. Since companies have economic goals to achieve, they must balance their activities in technical, economic and social terms. This balance results from the strong stakeholder pressure to ensure that technical, operational and organizational processes are safe, which should contribute to the smooth achievement of economic objectives. The digital transformation and progress in rail transport makes rail transport safety dependent on the safety of individual IT applications, which—combined into a single organism—must ensure the highest level of safe integration. Digitalization means that these applications, which can be divided into those that are directly related to railway traffic safety and those that are not directly related to railway traffic but which support these processes, together form the digital ecosystem of the railway system. A large number of systems created by these applications lead to the need to improve railway traffic safety management processes analyzed in terms of data and information transmission safety. The digitalization of the business models of railway enterprises and the digitalization of the railway system itself means that in general a multi-organization railway system consisting of the components of the railway system and digital business models is shaped. In some areas of activity, the boundaries between the railway system itself and the components of the digital business model of railway enterprises are even blurred. All this is based on the assumption that safety, especially in railway organizations, is a social factor that is part of a group of factors related to the concept of sustainable business models. This approach should consider digitalization processes while taking rail transport digital safety into account.

Methodological goals, which included:

1. Discussing key issues related to digital safety, its definitions, assumptions and practical implications in the rail transport sector.
2. Identifying digital safety management mechanisms in the context of the digital transformation of the business models of railway enterprises and railway organizational, social and technical systems.
3. Developing the conceptualization and operationalization of the digital safety model of rail transport.
4. Determining whether there are sufficient grounds for creating a new paradigm of digital safety management in rail transport have been achieved.

Digital safety management is emerging as a new paradigm for the management of the safety function and results from the need to rapidly digitalize the rail transport sector in terms of the railway system and the digital business models of railway enterprises. A model which is a set of commonly accepted assumptions in a given field, something that can be referred to, is described in the monograph. The paradigm of rail transport digital safety in this respect allows for the creation and organization of the fundamental knowledge in this thematic area and provides a platform for the implementation of utilitarian solutions, especially in the field of safety management mechanisms.

Key research questions, namely:

Question 1.Does the digital transformation of the business models of railway enterprises and systems have an impact on railway traffic safety?
Question 2.What digital safety management mechanisms determine technological progress?
Question 3.What are the key problems faced by rail transport safety managers in implementing innovative solutions for the digital economy?
Question 4.What are the key factors that determine aspects of cybersecurity in rail transport?
Question 5.What are the limitations of the different technologies and solutions of the digital economy in the context of rail transport safety?
Question 6.What determinants influence the treatment of rail transport safety as a social factor?

have been thoroughly analyzed in the monograph.

The answers to these questions have been given throughout the book and confirm the interdisciplinarity of the topic.

It should be pointed out that the subject discussed is highly complex and the issues raised in the monograph are numerous. As regards the bibliographic query, sources of scholarly literature, industry standards, thematic periodicals and European and national law were used in the work. This was the basis for identifying core assumptions for the construction of the mechanisms of digital safety in the business models of railway enterprises in management, social and technological terms.

As part of research into digital safety management mechanisms in the business models of railway enterprises, an analytical and evaluation study was created, the conclusions of which are summarized below:

1. Rail transport safety management mechanisms define a method based on a process approach to identify the key procedural rules that must be fulfilled to ensure the expected level of safety of a social and technical system such as the railway system.
2. Safety culture is an organizational tool which aims to maintain a constant focus on ensuring high reliability and the expected level of safety of the functioning of the railway system

3. The starting point in the context of complex technical systems, namely the railway system, will be the theory of the system of systems, as it is at the forefront of ensuring the integration of technical systems.
4. The effective management of technical, operational and organizational change in rail transport is a condition and the result of the verification of technical compatibility and safe integration. The standardization of interfaces between different types of equipment as part of the configuration of railway systems in investment processes, including the design, construction, renewal and modernization of railway systems, will be a core problem to be solved both in the technical but also organizational context of the project approach.
5. The digitalization of maintenance processes is important for the operation of railway vehicles and related safety management mechanisms. Prediction, which is crucial for the maintenance of railway vehicles based on digital solutions, allows for the determination of future states of objects through continuous measurement (including in real time), the analysis of the values of current and historical states, and the estimation of future values.
6. One of the core theories that are particularly relevant to the digitalization of the rail transport sector is the theory of cybernetics, understood as a transdisciplinary approach to control systems and the associated communication in both man and machine.
7. At present, as regards the understanding the digital business models of railway enterprises, the so-called extended business models are referred to in this respect, which combine functions at company level with interfaces at the level of the entire railway system, and all or part of the rail transport chain therein.
8. Safety can certainly be identified as an important social factor, which is the consequence of the social implications of the business model in rail transport.
9. Dynamic balance in rail transport is an important attribute of its existence in the organizational, management, economic, social and technological dimensions. The pursuit of a sustainable transport system throughout the supply chain is a fundamental need embedded in a modern understanding of the flows of goods and services, both locally and globally.
10. Linking rail transport and its mechanisms to, inter alia, the sharing economy, the circular economy, the Big Data economy, cloud computing, artificial intelligence, and the social media economy generates new added value, in both technological and social terms, for a wide range of stakeholders in the sector.
11. The flow of information in complex socio-technical systems, such as the railway system, is essential to ensure the expected level of rail traffic safety.
12. The digitalization of railway traffic control solutions is the most important aspect of the digital transformation of railways. Such control-command and signaling on-board and trackside subsystem solutions generate the need for digital transformation, including in other structural and functional subsystems such as infrastructure, energy, rolling stock, rail traffic, telematics, and supporting IT applications not directly related to railway traffic.
13. In the interoperable railway, a new dimension of the transport processes, the principles of railway infrastructure management and the mechanisms of the

operation of railway vehicles in rail transport are emerging. Digitalization is fundamental in this triad of three core areas shaping the rail transport system.

14. In rail transport, data and information are very often shared, which means that they are processed over time as part of their migration. Information strategies may consist of a single shared database or distributed heterogeneous federated repositories. A high level of information safety must be ensured in each model.
15. Digitalization is key to the increased accessibility, automation, communication and sustainable development of railway infrastructure. In this area, the solutions used should be safe.
16. Data migration in dynamic and complex technical and operational systems supported by organizational processes must be supported on the one hand by the appropriate selection of data processing tools, and on the other hand, forced by operational needs. However, decisions on the use of specific solutions must be strategic.
17. Cybersecurity is already a highly formalized concept based on legal regulations and standards. High levels of awareness of hazards, responsibility of managers and the level of training of railway system operators to respond to crisis situations are a condition for a proper approach to these problems. Protection against cyber-attacks of technical systems is currently one of the core issues for the construction of safety management systems of railway undertakings and infrastructure managers that meet the definition of operators of a critical service.
18. The level of application of the digital solutions of rail transport operators is very wide and could lead to the full digital transformation of the rail transport sector in the near future. Innovation in this sector is an inherent feature of its progress.
19. The widespread use of artificial intelligence determines the progress of the rail transport sector. The main obstacle to the creation of AI-based solutions is the complexity of the real world description.
20. Both safety-related and non-safety computer applications are used in the rail transport system. Together they form one technical ecosystem. Safety-related and non-safety-related telegrams and information are transmitted between applications. In any case, safety-related cryptographic techniques should be used to achieve the safety of data and information and in particular the safety of operations. Only a common strategy for managing data transmission processes from all IT applications involved in the functions of the railway system can ensure the expected level of digital safety.
21. The digital safety model is a coherent set of devices, technical systems and components supporting a safety management system, which uses digital solutions as a basis for maintaining the expected level of safety. The digital safety model is a combination of digital economy solutions embedded in individual components—the functions of the railway system. It is a set of components which shapes the configuration of the railway system, the protection of which against defects, failures, incidents, disasters, and any other irregularities is

built on a cybersecurity canvas based on technical, legal and organizational and normative solutions.

22. Rail transport safety management mechanisms should be based on and support the business model of railway enterprises; it can increasingly already be called a digital business model in many cases. This is due to the fact that most of the operations conducted by railway enterprises rely almost entirely on digital solutions. Safety management systems and digital technologies are then based on digital economy safety systems, including in particular electrical/electronic and programmable systems for which functional safety requirements are described in the relevant technical standards.
23. The safety triad proposed in the monograph consists of a set of three interconnected components, namely safety management mechanisms as a set of rules, procedures, rules and methods used to achieve the expected level of railway safety. The second components is the digitalization of processes related to rail transport activities, with a particular focus on digital safety. The final component of the triad is the socialization of the concept of railway traffic safety, which results from the expectations of stakeholders in terms of intolerance to railway incidents. At the center is the digital business model of railway enterprises, the components of which overlap with the entire area of the digital railway system. The level of integration of the digital business models of railway enterprises with the railway system is growing dynamically with progress in rail transport. It can be expected that in the future it will be difficult to unambiguously identify ontologically separated components of the digital business model from the digitalized railway system.
24. In the rail transport sector, in the field of two key research spaces that can be ontologically separated, namely safety and security, combining these concepts into a single model called 'Cyber Operational Safety' (COS) is proposed. Its assumptions should cover all risks arising from the use of computer applications directly related to safety and those that only support operational, maintenance and logistic processes in rail transport.

The conclusions presented result from the subject of research related to the subject of the monograph. The in-depth analysis allowed the authors to look at the issue broadly and to identify a number of conclusions, recommendations and observations which explain the knowledge gap in this area. The issue of digital safety management mechanisms and digital business models of railway enterprises in management, social and technological terms is a topic that is recent, modern, and requires a broad explanation.

The manufacturer's authorised representative in the EU is Springer Nature Customer Service Centre GmbH, Europaplatz 3, 69115 Heidelberg, Germany. If you have any concerns regarding our products, please contact ProductSafety@springernature.com

Printed and bound by CPI Group (UK) Ltd, Croydon, CR0 4YY
28/09/2026
02224094-0002